Mobil New Zealand Nature Series

Alpine Plants
of New Zealand

Mobil New Zealand Nature Series
Alpine Plants
of New Zealand

Lawrie Metcalf

Front cover: *Gentiana bellidifolia* var. *australis*
Title page: *Leucopogon fraseri*

Published by Reed Books, a division of Reed Publishing (NZ) Ltd, 39 Rawene Road, Birkenhead, Auckland. Associated companies, branches and representatives throughout the world.

First published 1996
Text and photographs © Lawrie Metcalf 1996

This book is copyright. Except for the purpose of fair reviewing, no part of this publication may be reproduced or transmitted in any form or by any means, electronic or mechanical, including photocopying, recording, or any information storage and retrieval system, without permission in writing from the publisher. Infringers of copyright render themselves liable to prosecution.

ISBN 0 7900 0525 5

Printed in Hong Kong By South China Printing Co.Ltd
Colour Separations by Image Centre Ltd New Zealand.

Contents

Introduction · vii

Lycopodium fastigiatum · 1
Blechnum pennamarina · 2
Podocarpus nivalis · 3
Lepidothamnus laxifolius · 4
Ranunculus insignis · 5
Ranunculus lyallii · 6
Notothlaspi rosulatum · 7
Viola lyallii · 8
Geranium sessiliflorum · 9
Epilobium macropus · 10
Pimelea oreophila · 11
Acaena caesiiglauca · 12
Geum cockaynei · 13
Carmichaelia monroi · 14
Anisotome haastii · 15
Gingidia montana · 16
Aciphylla colensoi · 17
Aciphylla monroi · 18
Gaultheria depressa · 19
Cyathodes colensoi · 20
Pentachondra pumila · 21
Leucopogon fraseri · 22
Dracophyllum longifolium · 23
Coprosma cheesemanii · 24
Celmisia semicordata · 25

Celmisia sessiliflora	26
Leptinella atrata	27
Raoulia bryoides	28
Raoulia subsericea	29
Leucogenes grandiceps	30
Helichrysum bellidioides	31
Brachyglottis bellidioides	32
Brachyglottis bidwillii	33
Cassinia vauvilliersii	34
Gentiana bellidifolia	35
Wahlenbergia albomarginata	36
Wahlenbergia laxa	36
Donatia novae-zelandiae	37
Phyllachne colensoi	38
Pratia angulata	39
Euphrasia revoluta	40
Ourisia macrocarpa	41
Ourisia macrophylla ssp. *robusta*	42
Parahebe lyallii	43
Hebe macrantha	44
Astelia nervosa	45
Bulbinella hookeri	46
Chionochloa rubra	47
Further reading on this subject	xciv
Index of common names	xcv
Index of scientific names	xcvi

Introduction

New Zealand has a rich alpine vegetation, particularly in the South Island, although there is still quite a good representation of alpine plants on the mountains of the North Island and also on Stewart Island.

Out of a total native flora of more than 2000 species approximately 600 are true alpines. In addition, approximately another 350 species occur in both lowland and mountain areas so that the high mountain flora represents almost 50 per cent of our total flora. Of the high mountain plants some 420 are confined to the mountains of the South Island.

Alpine plants are generally considered to be those which grow in or are confined to the alpine zone or high country, which occurs above the tree-line in the three main islands. The altitude of the tree-line varies, being highest in the north and lowest in the south. In the North Island it ranges from about 1450–1500 metres in the upper central area and descends to about 1200 metres on the Tararua Range in the south. In the South Island it ranges from about 1200 metres on the northernmost mountains to about 900 metres in the far south.

The northernmost limit for native alpine plants is on Te Moehau (840 metres) at the northern tip of the Coromandel Peninsula where a patch of open ground supports an interesting array of alpine species. The next area where alpine plants occur is about 270 kilometres to the south-east on Mt Hikurangi, near East Cape. From there, southwards, the range of species steadily increases until reaching the headquarters of our alpine flora, in the mountains of the South Island.

The type of plant cover, climate and changes to the vegetation are used to define the altitudinal zones mentioned in this book. They are: lowland, montane, subalpine and alpine. The alpine zone is further divided into low alpine, high alpine and nival. Each of these zones extends over 300–500 metres of altitude although, as already explained, their actual elevations will vary according to their latitude from north to south. They also vary on individual mountains, or mountain ranges, according to the aspect of the slope.

The lowland zone commences at sea level and it is not until the subalpine zone is reached that alpine vegetation usually commences to appear. The majority of species grow in the alpine zone and beyond that is the permanent snow line or nival zone, where virtually no plants grow.

The alpine zone supports a number of vegetational types which are referred to in this book; the main ones being:

1. **Mixed snow tussock-scrub**: occurs just above the tree-line and usually comprises one or more species of the larger snow grasses (*Chionochloa*) and taller alpine shrubs.
2. **Snow tussock-herbfield**: mainly occuring in high rainfall areas, this association gradually takes over from the mixed snow tussock-scrub. It is dominated by one of the larger species of snow grass intermixed with larger herbs such as *Celmisia*, *Aciphylla* and *Astelia*.
3. **Herbfield**: the upper limits of the snow tussock-herbfield gradually merge into the herbfield proper. Herbfields are characterised by a wide range of large, medium and small herbs. The large snow tussocks do not dominate the vegetation either because conditions are unsuitable, or they have been kept in check by browsing animals. There is a wider range of species in high rainfall regions than on the drier eastern ranges, and the cooler south-facing slopes are also richer than the warmer north-facing ones.
4. **Snow tussock grassland**: this is a characteristic feature of the drier alpine areas of the South Island, east of the main divide. It is usually dominated by one of the larger snow tussocks such as *Chionochloa rigida* or *C. macra*.
5. **Bogs**: these are areas of poor drainage, usually in depressions, but sometimes on flatter areas such as the tops of alpine passes. Often they are associated with tarns and small pools. When boggy areas occur on slopes they are known as flushes or seepages. Bogs are more common in higher rainfall areas. Typical bog plants are small rushes,

sedges and similar plants, and cushion plants such as *Donatia*, *Phyllachne* and *Lepidothamnus laxifolius*.
6. **Fellfield**: towards the upper limit of the alpine zone, plant cover often tends to be sparser as the ground becomes stonier, with less soil and fewer plants. Fellfields are usually more or less stable.
7. **Scree**: screes are areas of loose stones, usually more common on the drier greywacke mountains of the South Island. The more ancient screes on gentler slopes are relatively stable, but can be quite mobile on steeper slopes. Many of the screes seen today are not natural and have been induced over the past 120 years or so by continual burning and grazing of the plant cover. Most scree plants are highly specialised.
8. **Cushion vegetation**: a very dwarfed, tundra-like vegetation most common on the rolling summits of the Central Otago ranges. It comprises dwarf species including *Dracophyllum*, *Phyllachne* and *Anisotome*.
9. **Snowbank vegetation**: near the summits of some ranges where there are depressions sheltered from the prevailing wind, banks of snow accumulate and often remain until about late summer. Plants in snowbank areas often have very short growing seasons, depending on the depth of the snow and how quickly it thaws.

With such a rich alpine flora it is not intended, nor is it possible, to make this book a comprehensive guide. Its main purpose is to provide examples of the main genera likely to be seen in the mountains so that people can become more familiar with them. Many of the plants illustrated do not have common names and no attempt has been made to coin them where none exists.

I hope that readers find this guide useful and informative, and that it encourages further interest in New Zealand's alpine flora.

Lawrie Metcalf

▲*Lycopodium fastigiatum*
Lewis Pass, December 1995

▼*Lycopodium scariosum*
Upper Wairau River, March 1996

1 *Lycopodium fastigiatum*
Alpine Club-moss

Family LYCOPODIACEAE

Description
- A creeping plant with branched, underground main stems. Aerial stems erect and much-branched with branches upturned. In sheltered situations, may be up to 25 cm tall; in exposed places or at higher altitudes it can be dwarfed with the branchlets almost prostrate on the ground. Colour of the plant is usually a subdued bronzy green or sometimes quite orange.
- *Leaves*: Narrow or linear, and spirally arranged around the branchlets.
- *Cones*: The erect, spore-producing cones are up to 7 cm long on long stalks; solitary, in pairs or several together.

Distribution & Habitat
- North, South and Stewart Islands in montane to high-alpine regions, but descending to lowland areas in the far south. Sea level to 2000 metres.
- Commonly occurs in open or more or less open sites, particularly in subalpine scrub, alpine herbfield, grasslands and boggy areas. At higher altitudes it may be less common and is often quite dwarfed and inconspicuous.

Notes
- *Related species*: The creeping plant *L. scariosum* has a similar range, except that it is less common north of Rotorua and along the drier eastern areas of the North and South Islands. It also occurs in similar habitats. It differs from *L. fastigiatum* in its bright green colour and the broader leaves lying more or less in the one plane so that the branchlets have a flattened appearance. The spore-producing cones are produced only singly.

2 *Blechnum pennamarina*
Alpine Hardfern

Family BLECHNACEAE

Description
- The rhizomes (underground spreading stems) creep to make the plant often quite wide-spreading.
- *Fronds*: Numerous, tufted at the tips of short branches or distributed along the rhizome. There are two kinds: sterile and fertile; both long and narrow on wiry stalks.

    Sterile fronds 3–15 cm long, 1–1.5 cm broad. Upper surface green to bronzy green, segments (pinnules) closely placed along the midrib and attached by their broad bases. Newly emerged sterile fronds often bronze or quite reddish.

    Fertile fronds usually erect, shorter and narrower, blackish when young and becoming brownish at maturity. The narrow pinnules often point downwards.

Distribution & Habitat
- North, South and Stewart Islands. Local from Northland to the Bay of Plenty, but widespread from there southwards. Sea level to 2000 metres.
- Occurs from lowland to high alpine areas, being found in scrub, tussock grassland, herbfield, fellfield and on old moraines. In drier mountain areas, often confined to more shady situations.

Notes
- *Identification*: Easily recognised by its narrow fronds, the segments being attached to the midrib by their broad bases; and thin, wiry stalks. At higher altitudes it is often considerably dwarfed and not always readily distinguishable amongst other vegetation.

◀ *Blechnum pennamarina*
Jack's Pass, December 1995

▲ *Podocarpus nivalis*
Upper Wairau Valley, March 1996

▼ *Podocarpus nivalis,* showing pollen cones, Otira Valley, January 1996

3

Podocarpus nivalis
Mountain Totara / Snow Totara

Family PODOCARPACEAE

Description
- A prostrate, sprawling or erect small shrub up to 2 metres tall. Branches often wide-spreading and rooting into the ground.
- *Leaves*: Closely set on the branchlets and more or less spirally arranged, 8–16 mm by 1.5–3 mm, bronze to deep green or yellowish to deep green. Leaf tips have small pungent points.
- *Cones*: Plants are unisexual: on the male the pollen cones are obvious, 7–25 mm long and usually produced during early summer; on the female the fruits ripen during late summer and autumn.
- *Seed*: The short foot-stalk becomes swollen and fleshy, and is orange or scarlet when ripe with a small, green, nutlike seed sitting on top.

Distribution & Habitat
- North and South Islands from Te Moehau on the Coromandel Peninsula southwards. Inclined to be local in the North Island, but is found throughout most mountain areas of the South Island. Usually from 700–1500 metres, but descends to near sea level in parts of South Westland.
- Occurs commonly in subalpine scrub, mixed snow tussock-scrub and amongst the shrubby vegetation that colonises old moraines. Not uncommon in the higher altitude alpine forests.

Notes
- *Identification*: The mountain totara varies in size and habit and in some areas may be completely prostrate to form springy mats many metres across. In other areas it forms an erect or more or less erect shrub up to a metre or more in height.
- *Other*: On the drier mountains east of the main divide its spreading and rooting habit sometimes helps to stabilise slopes of loose rock debris.

▲ *Lepidothamnus laxifolius*
Lake Sylvester, January 1996

4. *Lepidothamnus laxifolius*
Pygmy Pine

Family PODOCARPACEAE

Description
- A usually prostrate, small shrub with slender trailing branches up to 1 metre or more long.
- *Leaves*: On juvenile plants the narrow leaves are 5–8 mm long and stand out from the branchlets; on older plants the leaves are reduced to 1.5–3 mm in length and are appressed to the branchlets. Branchlets often show a mixture of different leaf forms.
- *Cones*: Mostly plants are unisexual. The male pollen cones are produced during early summer and the fruits ripen from late summer to late autumn.
- *Seed*: The swollen and fleshy footstalk turns bright crimson and has a small blackish nutlike seed sitting on top.

Distribution & Habitat
- North Island from Mts Tongariro and Ruapehu southwards. Common throughout the South and Stewart Islands. Usually 760–1500 metres but descends to near sea level on Stewart Island.
- Occurs mainly in boggy or poorly drained sites in subalpine and low alpine areas. Common in sphagnum and cushion bogs as well as snow tussock-herbfield.

Notes
- *Identification*: Foliage colour varies considerably and can be anything from a blue-grey to green or bronze, or even slightly purplish. Much depends on the degree of exposure and the time of the year.
- *Flowering*: While it is said to be unisexual, plants may sometimes have both male and female flowers on them.

▲ *Ranunculus insignis*
Rainbow Ski Area, December 1955

5 *Ranunculus insignis*
Korikori

Family RANUNCULACEAE

Description
- This species varies from a robust herb 50–60 cm tall, to a smaller tufted plant no more than 10 cm or so tall.
- *Leaves*: Heart- or kidney-shaped, thick and leathery with brown hairs fringing their toothed margins. On the larger forms they are 15–22 cm in diameter, while on small forms they may be no more than 2–4 cm in diameter.
- *Flowers*: The flowering stem is branched (may be unbranched and with only a single flower in small forms) and carries numerous bright yellow flowers, 2–5 cm in diameter. Usually five to seven petals, but may be up to 12 on plants in some areas of Marlborough.

Distribution & Habitat
- North and South Islands in subalpine to low alpine regions: North Island from Mount Hikurangi southwards; South Island from Nelson and Marlborough to about mid Canterbury. 700–1800 metres. (Along parts of the Kaikoura coast it occurs almost down to sea level.)
- Occurs in shady areas of grassland, herbfield, subalpine shrubland, on sheltered bluffs and rock outcrops.

Notes
- *Identification*: With its bold, green leaves and bright yellow flowers it is a handsome and easily recognised species, athough, as currently understood, the species is an amalgam of several forms ranging from the larger plants, commonly regarded as *R. insignis*, to smaller plants formerly regarded as distinct species. The smallest forms generally occur in the southern part of its range.
- *Flowering*: October to December.

▲ *Ranunculus lyallii*
Arthur's Pass, November 1955

▼ *Ranunculus lyallii* foliage,
Otira Valley, January 1996

6 *Ranunculus lyallii*
Mountain Lily / Mt Cook Lily / Mountain Buttercup / Shepherd's Lily

Family RANUNCULACEAE

Description
- A robust herb, usually 60–75 cm tall, but in very good situations may be up to 1 metre.
- *Leaves*: Saucer-shaped, 15–40 cm in diameter, deep shining green, thick and rather leathery. Leaf stalks long and attached to the centre of the blade.
- *Flowers*: Numerous, on branched stems; 5–8 cm in diameter, white with a green cone-like centre surrounded by yellow stamens.

Distribution & Habitat
- South and Stewart Islands. About and west of the main divide, from Mt Buckland near Westport, southwards. Rare on Mt Anglem on Stewart Island. 700–1500 metres.
- Subalpine to low alpine throughout most of the wet mountain areas, particularly from Arthur's Pass southwards. Usually in snow tussock-herbfield, alongside streams, in wet hollows and flushes, and on rock bluffs and faces.

Notes
- *Habitat*: In some areas introduced browsing animals have virtually restricted it to inaccessible habitats such as rock bluffs.
- *Flowering*: This species is justly acclaimed as one of the most magnificent buttercups in the world. Depending on altitude it usually flowers from late November or early December to about mid January.

▲ *Notothlaspi rosulatum*
Island Pass, January 1996

7 *Notothlaspi rosulatum*
Penwiper

Family BRASSICACEAE

Description
- Usually forms a flattened rosette of leaves, up to about 8 cm across, arising from a single taproot.
- *Leaves*: Grey, fleshy, and arranged in overlapping layers, those of each layer becoming progressively smaller.
- *Flowers*: From the centre of the rosette a cone-shaped flowerhead, up to 25 cm tall, arises. It bears numerous creamy white, sweetly scented flowers.
- *Seed*: After flowering the broadly winged seed pods are produced: up to 2.5 cm long and more or less heart-shaped.

Distribution & Habitat
- South Island, on the drier greywacke mountains east of the main divide, from Marlborough and eastern Nelson to North Otago. 800–1800 metres.
- Confined to shingle screes or frost-eroded areas that are predominantly stony. Usually grows on the finer and more stable screes rather than those that are coarse and more unstable.

Notes
- *Identification*: When not in flower the grey foliage blends with the surrounding stones, making it difficult to locate.
- *Flowering*: Between November and January. The plant takes from two to three years to reach maturity and after flowering dies.
- *Name*: Its common name is derived from its resemblance to the old-fashioned cloth penwipers used in the 19th century.

8 *Viola lyallii*
Native Violet

Family VIOLACEAE

Description
- A tufted plant with sometimes wide-spreading stems.
- *Leaves*: Usually, but not always, distinctly heart-shaped; margins have rounded, shallow teeth.
- *Flowers*: On short stalks or up to about 8 cm long, white with purple veins on the lower petals and sometimes yellow in the centre.

Distribution & Habitat
- North and South Islands from about Kaitaia southwards. Sea level to 1800 metres.
- Usually in lowland to subalpine localities, in damp or moist places in grasslands, shrubland-grassland and rocky habitats.

Notes
- *Identification*: Plant varies from quite diminutive at higher altitudes, where it is often not easily noticed amongst other vegetation, to larger and more conspicuous at lower altitudes.
- *Related species*: Very similar to *V. cunninghamii* which mainly differs in its shorter stems and the leaves not being heart-shaped. Both species occur in similar habitats.
- *Flowering*: Between November and January.

▼ *Viola lyallii*
Old Man Range, December 1995

9 *Geranium sessiliflorum*

Family GERANIACEAE

Description
- Usually a rather small plant arising from a stout, central rootstock.
- *Leaves*: Often tightly clustered around the rootstock in the form of a rosette and flattened onto the ground. The leaf blade is 1.5–2 cm in diameter, kidney-shaped and the margins are cut into five to seven lobes which in turn are segmented.
- *Flowers*: The flower stalks are single-flowered and often so short that the flowers nestle amongst the leaves. Flowers white, about 1 cm in diameter.

Distribution & Habitat
- North, South and Stewart Islands from about the Waikato southwards. Coastal to high alpine areas. Sea level to 1700 metres.
- More common in lowland to subalpine grasslands, stabilised riverbeds and terraces, and similar stony areas. In the alpine zone it usually occurs in well-drained, open, stony sites, particularly those depleted by frost or browsing animals.

Notes
- *Identification*: The foliage colour can vary from green to a deep purplish-bronze, both colour forms often occurring in the one area. Plants growing in dry stony areas are often much stunted, whereas those in slightly better situations grow rather more vigorously.
- *Flowering*: Between November and January.

▼ ***Geranium sessiliflorum*** **Lake Tennyson, January 1996**

▲ *Epilobium macropus*
Otira Valley, January 1996

10 *Epilobium macropus*
Willow Herb

Family ONAGRACEAE

Description
- A low-growing herb with slender stems up to 25 cm long. The lower portions of the stems creep and root, while towards their tips they turn upwards. Upper third of the stem usually purplish.
- *Leaves*: In pairs on short, slender stalks, the pairs rather widely spaced. Leaf blade 10–15 mm by 4–10 mm, ovate (widest below middle) to oval, margins usually slightly toothed.
- *Flowers*: Produced along the upper part of the branchlets, white, 9–12 mm in diameter.
- *Seed*: Seed pods or capsules 4–5 cm long.

Distribution & Habitat
- North and South Islands where it is widely spread from the central North Island southwards. Quite widespread in montane to low alpine areas. 600–1500 metres.
- Most common in permanently wet places or in slow, shallow streams.

Notes
- *Identification*: A large genus of which some 50 species are native to New Zealand. Easily recognised by its flowers having four petals, each of which is deeply notched at the apex. The flowers appear to sit on top of a long and rather thick stalk, which is really the undeveloped seed capsule. Also, the leaves are in opposite pairs (each succeeding pair is turned at right angles to the previous pair).
- *Flowering*: Between November and January.

▲ *Pimelea oreophila*
Old Man Range, December 1995

11 *Pimelea oreophila*

Family THYMELAEACEAE

Description
- A prostrate to sprawling, rather sparingly branched, small shrub with brownish stems and branchlets. Plants usually about 30–40 cm in diameter.
- *Leaves*: 4–8 mm long by about 2–4 mm wide, upper surface usually greyish-green, margin reddish with scattered hairs around it.
- *Flowers*: White, clustered at the tips of the branchlets, about 5 mm in diameter.
- *Fruits*: Fleshy, about 4 mm long, orange-scarlet in colour.

Distribution & Habitat
- North Island on the Volcanic Plateau, Kaweka and Ruahine Ranges, and throughout most mountain regions of the South Island. In montane to high alpine areas. 500–2000 metres.
- Frequently common in tussock grasslands and herbfields, but also occurs in sheltered sites in the fellfield.

Notes
- *Identification*: Probably the most commonly encountered *Pimelea* in most mountain regions. When not in flower or fruit it is often inconspicuous. Superficially it resembles a *Hebe* but is easily distinguished by its tough stems which have fibrous bark and do not break as easily as those of *Hebe*. Its fleshy fruits also help to distinguish it from *Hebe* which has small, dry capsules.
- *Flowering*: Between November and January.

▲ *Acaena caesiiglauca*
Dunedin Botanic Gardens, December 1991

▼ *Acaena glabra* flowers and young seed heads
Mt St Patrick, January 1996

12 *Acaena caesiiglauca*
Biddy Biddy

Family ROSACEAE

Description
- A prostrate creeping herb, often quite wide-spreading, with the stems rooting into the ground, and the branchlets ascending. Most parts of the plant are densely covered with soft, white hairs.
- *Leaves*: Distinctly bluish- or greyish-green, 3.5–5 cm long with three to four pairs of leaflets and a larger terminal leaflet, all of which are toothed.
- *Flowers*: The brownish-green flower stalks are up to 15 cm long; flowering heads are about 1 cm in diameter. Each head comprises numerous small, white flowers.
- *Seed*: After the flowers have been pollinated the characteristic barbed spines develop on each seed. Seed heads turn brown when ripe and are about 2 cm in diameter.

Distribution & Habitat
- South Island, mainly in the drier mountains east of the main divide. 600–1500 metres.
- Usually widespread in grasslands in montane to low alpine regions.

Notes
- *Name*: The common name of biddy biddy (also spelt bidibidi) is derived from the Maori 'piripiri', and is often shortened to biddy bid.
- *Seed dispersal*: Most species of *Acaena* are characterised by the spines of the seed heads being barbed at their tips. When anything suitable, such as sheep's wool, woollen clothing, socks and other fibrous materials, comes into contact with them the barbs hook onto the fibres and the seeds are pulled free. This enables the species to colonise new areas. One or two species appear to have successfully dispensed with this aid to seed dispersal. With *A. inermis* and *A. glabra* the seed heads have no spines, while the seed heads of *A. microphylla* have spines but they are without barbs. This does not appear to be any impediment to the natural dispersal of those species.

13 *Geum cockaynei* (*G. parviflorum*)

Family ROSACEAE

Description
- A tufted herb forming rosettes of bright green leaves, 10–30 cm tall at flowering.
- *Leaves*: 6–15 cm long with a very large terminal leaflet and up to 15 pairs (usually less) of quite small leaflets along the midrib. Margin of the terminal leaflet toothed.
- *Flowers*: Flower stem hairy, and sometimes well-branched. Flowers white, up to 1.5 cm in diameter.

Distribution & Habitat
- North and South Islands from about Mt Hikurangi and the Ruahine Range southwards. 800–1700 metres.
- Mainly in low alpine to subalpine grasslands, herbfields and on rock faces. Absent from the central volcanic plateau, Mt Taranaki and Central Otago.

Notes
- *Identification*: Easily recognised by its bright green leaves, each of which has a very large terminal leaflet. The branched and hairy flower stems, and white flowers also help to identify it. Although it grows in a variety of situations, it is perhaps most noticeable when it grows on bluffs and rock faces.
- *Flowering*: Between November and January.
- *Name*: Recent research has meant that the familiar name of *G. parviflorum* has now had to be replaced by *G. cockaynei*.

▼ *Geum cockaynei*, Queen's Park, Invercargill, December 1991

Carmichaelia monroi

Family FABACEAE

Description
- A dwarf, much-branched shrub, arising from a single taproot and forming hard, flattish or slightly hummocky patches up to 20 cm in diameter.
- *Branchlets*: Hard and rigid, flattened and 2–4 mm wide.
- *Flowers*: Pinkish- or reddish-purple, produced in clusters of two to five from small notches along the sides of the branchlets.

Distribution & Habitat
- South Island: in the drier mountain areas east of the main divide from Marlborough to the Pisa Range in Central Otago. 800–1500 metres.
- Montane to low alpine areas, usually in open habitats in sparse or depleted low tussock grassland where the ground is stony or rocky and well-drained.

Notes
- *Identification*: In spite of the hard nature of its branchlets this plant is frequently browsed by hares so that it seldom develops to its true appearance; the tips of the branchlets looking as though they have all been cut off. The adult is completely leafless and only very young seedlings have leaves.
- *Flowering*: November or December.

▼ *Carmichaelia monroi*
Queen's Park, Invercargill, December 1991

15 *Anisotome haastii*

Family APIACEAE

Description
- A large tufted herb with deep-green, fern-like leaves. Where conditions are suitable it will form a large clump up to 30 cm across, but more often it is rather smaller.
- *Leaves*: 10–25 cm long by 5–12 cm broad; finely divided into small, narrow segments.
- *Flowers*: The flowering stem may be up to 60 cm tall. Flowers white, small and numerous, produced in compound heads.

Distribution & Habitat
- South Island in subalpine to low-alpine regions, in high rainfall areas throughout. Rare or absent from the drier mountain areas such as Marlborough and Central Otago. 600–1520 metres.
- Usually in herbfields, tussock-herbfields, fellfields and on rock bluffs.

Notes
- *Seed*: The male and female flowers are produced on separate plants so that seed is only formed on the female plants.
- *Habitat*: In some areas browsing by introduced mammals has restricted this species to inaccessible places such as rock bluffs, but it is again becoming more common where browsing pressure has been reduced.
- *Flowering*: Between October and February.

◀ **Anisotome haastii**
Otira Valley, January 1996

▲ *Gingidia montana*
Black Birch Range, February 1996

16 *Gingidia montana*
New Zealand Angelica

Family APIACEAE

Description
- Usually a rather stout, tufted, very aromatic herb with a deep and fleshy taproot.
- *Leaves*: Mostly forming a rosette; up to 40 cm long and 10 cm wide, with five to ten pairs of broad leaflets. Bright, shining green above and somewhat glaucous (greyish-green) beneath; margins finely but distinctly toothed.
- *Flowers*: Flowering stems up to 50 cm tall with compound flower heads, up to 10 cm in diameter, at the top. Flowers small and white, giving the flower heads a foamy appearance.

Distribution & Habitat
- North and South Islands from about Kawhia and Lake Taupo southwards. In the North Island sporadic and generally absent from alpine areas; common in lowland and alpine areas throughout the South Island except for South Canterbury, north and Central Otago, western Southland and southern Fiordland. Lowland to low-alpine areas to 1500 metres.
- Usually grows in moist, open sites in forest, scrub snow tussock-herbfield, on stream banks and rock faces.

Notes
- *Habitat*: While formerly very common, this plant is very palatable to browsing animals and is now frequently confined to inaccessible rock faces where they cannot easily reach it. Where animal numbers have been greatly reduced it can once again be found in its former habitats.
- *Flowers*: Generally, the male and female flowers occur on separate plants although, occasionally, individual plants may produce some flowers of the opposite sex.
- *Flowering*: Between November and January.

▲ *Aciphylla colensoi*
Moa Park, December 1995

▼ *Aciphylla colensoi,* female,
Moa Park, December 1995

17 *Aciphylla colensoi*
Speargrass / Spaniard

Family APIACEAE

Description
- A large herb arising from a strong taproot and forming single or multiple rosettes of very sharp, spine-tipped leaves. The clumps may be up to 90 cm in diameter and 40–50 cm high.
- *Leaves*: Rigid, divided into long and narrow segments, all pointing forwards and outwards in all directions. Usually leaves are green or greyish-green with prominent orange or reddish midribs.
- *Flowers*: Small, yellowish flowers are produced in dense clusters along strong stems up to a metre or so tall. Long, narrow spines project out from amongst the flowers to give the appearance of quite a formidable protection.

Distribution & Habitat
- North and South Islands from Mount Hikurangi to mid Canterbury. Widespread in subalpine to low-alpine areas. 900–1500 metres.
- Often prominent in subalpine scrub, mixed snow tussock-scrub, herbfields and in grasslands. Usually in moister situations.

Notes
- *Identification*: This fierce-looking plant is particularly conspicuous at flowering time when the bright orange flowering stems stand out from afar. With all species of speargrass the male and female flowers are on separate plants with the male usually being the more showy.
- *Related species*: There are some 40 species of *Aciphylla*, most of which occur in alpine regions. They range from quite small species no more than a few centimetres tall to those up to 2 metres or more tall at flowering.
- *Other*: Some have thought the spiny nature of speargrasses to have been a protection against the browsing activities of the now extinct moa, but in view of the ease with which introduced animals such as sheep, rabbits and hares browse on them the spines are unlikely to have afforded much protection against browsing moa. A more likely explanation is that the spiny nature of the plant is a response to habitat, particularly as a means of moderating the effects of dessicating winds.

▲ *Aciphylla monroi*
Rainbow Ski Area, January 1996

▼ *Aciphylla similis*, female plant in fruit, Otira Valley, January 1996

18 *Aciphylla monroi*

Family APIACEAE

Description
- A small, tufted herb, usually yellowish-green in colour.
- *Leaves*: Usually rather crowded, with one terminal leaflet and two to four pairs of sharply pointed leaflets or occasionally with up to six or eight pairs.
- *Flowers*: Flowering stems 15–20 cm long, flowers whitish, numerous.

Distribution & Habitat
- South Island from Nelson and Marlborough to about mid Canterbury. 1100–1700 metres.
- Fairly common on rock outcrops, in snow tussock-herbfield and often in partly eroded and open snow tussock grassland. Perhaps rather more common on the drier eastern mountains.

Notes
- *Related species*: *Aciphylla monroi* is only likely to be confused with *A. similis* which occurs from just north of the Lewis Pass to South Canterbury and the mountains around the southern end of Lake Wakatipu. *A. similis* has four to ten pairs of leaflets which are wider spreading than those of *A. monroi*. Also its flowering stems are up to 40 cm long and it usually occurs in the wetter mountain regions.
- *Flowering*: Between December and February.

▲ *Gaultheria depressa* var. *novae-zelandiae*
Upper Wairau River, March 1996

▼ *Gaultheria depressa* var. *novae-zelandiae*
Lake Sylvester, January 1996

19 *Gaultheria depressa* Snowberry

Family ERICACEAE

Description
- A small, much-branched, low-growing shrub, usually with prostrate or trailing branches that root into the ground to form matted patches 20–30 cm across.
- *Leaves*: 5–10 mm by 4–8 mm, rounded with small teeth around the margins, and deep green above.
- *Flowers*: White, 2–3 mm long, urn-shaped and often more or less concealed amongst the foliage.
- *Fruits*: Large and fleshy, conspicuous; up to 1.5 cm in diameter. Vary in colour from white to pink or red.

This species exists in two distinct **forms**.
- *Typical form*: Has short, bristle-like hairs around the margins of its almost round leaves.
- var. *novae-zelandiae*: Has somewhat narrower and more pointed leaves without the bristles around the margins.

Distribution & Habitat
- *Typical form*: North and South Islands in lowland to low-alpine areas of the southern Tararua Range in the North Island and then along the main divide of the South Island to South Otago. 500–1500 metres.
- var. *novae-zelandiae*: Widespread in lowland to low-alpine areas from the Volcanic Plateau southwards. To 1500 metres.
- Both occur in similar habitats in subalpine forest and scrub, open areas in snow-tussock grassland, herbfields, rock outcrops and road-banks.

Notes
- *Fruits*: White-fruited forms often appear to be the most common in many areas, particularly in more open country, whereas in forest or scrub areas pink or red fruits are perhaps more often seen. The fruits are edible, although not overly flavoursome. Early settlers, in southern districts, used to gather the fruits for making snowberry pies. Fruits are usually on plants between January and April.

▲ *Cyathodes colensoi*
Upper Wairau River, March 1996

▼ *Cyathodes colensoi,* pink fruited form,
Upper Wairau Valley, March 1996

20 *Cyathodes colensoi*

Family EPACRIDACEAE

Description
- A small, low-growing shrub up to about 40 cm tall by 60 cm across, but occasionally much wider spreading and then forming broad flattened patches.
- *Leaves*: Small, rigid; distinctively characteristic shape, being oblong and then abruptly narrowed to a minute point. Dark green above, pale greyish beneath with several distinct veins. The margins are rolled slightly downwards and inwards.
- *Flowers*: Small, white, tubular, about 4–5 mm long and in clusters of two to five at the tips of the branchlets. The five spreading lobes at the mouth of the tube are distinctly hairy.
- *Fruits*: 4–5 mm in diameter, and range from red to pink to white, on different plants.

Distribution & Habitat
- North and South Islands from the Volcanic Plateau, Kaimanawa, Ruahine and Tararua Ranges southwards to Otago and northern Southland in montane to low-alpine regions. 600–1600 metres. In the South Island it is confined to the drier ranges east of the main divide from eastern Nelson and Marlborough southwards. It is inclined to be local in Otago and is rare in northern Southland.
- Usually in well-drained situations in mixed tussock-scrub, snow tussock grassland and on rock outcrops.

Notes
- *Scent*: The flowers of this plant have a very distinctive honey scent and sometimes it may be detected before the plant is seen.
- *Fruits*: The fruits are quite long-lasting and it is not uncommon to observe plants that are both in flower and fruit. Although the flowers appear to have both male and female parts, generally they function as either male or female so that those plants which are functionally male do not usually bear fruits.
- *Flowering*: Between November and January.

▲ *Pentachondra pumila*
Mt Peel (Nelson), January 1996

21 *Pentachondra pumila*

Family EPACRIDACEAE

Description

- A dwarf, much-branched, creeping shrub which forms dense patches or low hummocks of deep- or bluish-green foliage. The patches are usually 2–4 cm thick and vary from 30–40 cm or more across.
- *Leaves*: Small and hard, 3–5 mm by 1–2 mm, deep green above and slightly paler beneath.
- *Flowers*: Small, whitish, honey-scented, borne singly near the tips of the branchlets; similar to those of *Cyathodes colensoi*.
- *Fruits*: Bright red, berry-like, 5–6 mm in diameter, hollow inside and with five or more small nutlets.

Distribution & Habitat

- North, South and Stewart Islands in subalpine to low-alpine areas from the Coromandel Peninsula southwards. Very common in the South and Stewart Islands, but rare north of East Cape in the North Island. 600–1500 metres.
- Occurs in cushion bogs, open snow tussock grasslands, herb-fields and herb moor. Frequently grows in exposed or rocky sites or in poorly drained peaty areas.

Notes

- *Fruits*: The fruits take two seasons to ripen, being quite small and green by autumn and maturing during the following summer.
- *Identification*: The hollow fruits distinguish this species from other plants of a similar appearance.
- *Flowering*: Between November and January, but occasional flowers may be seen at other times.

▲ *Leucopogon fraseri*
Cobb Valley, January 1996

▼ *Leucopogon fraseri*
Queen's Park, Invercargill, November 1991

22 *Leucopogon fraseri*
Patotara

Family EPACRIDACEAE

Description
- A small, low-growing shrub with creeping, underground stems, often forming quite extensive patches. The wiry stems are usually 5–15 cm tall and clothed for much of their length with foliage.
- *Leaves*: 4–9 mm long and 1–3 mm wide, quite hard of texture and the tip is abruptly narrowed to a fine but pungent point. Their upper surface is deep green, greyish, or yellowish- to bronzy green.
- *Flowers*: White, produced from the leaf axils along the stem, strongly honey-scented, 1–1.3 cm long, with five distinctly bearded lobes at the mouth of the tube.
- *Fruits*: Orange, 8–9 mm long.

Distribution & Habitat
- North, South and Stewart Islands in coastal to low-alpine regions throughout. Sea level to 1600 metres.
- Usually common in dry places in coastal dune hollows, low tussock grasslands, rocky places and fellfields.

Notes
- *Identification*: The leaves can be quite prickly to the touch, which is often the first indication of its presence amongst other low vegetation.
- *Fruits*: Edible and have a sweetish if somewhat resinous taste. *L. fraseri* often has a long flowering season and both flowers and fruit can be seen on plants at the one time.

23 *Dracophyllum longifolium*
Grass tree / Inanga / Inaka

Family EPACRIDACEAE

Description
- An erect shrub or small tree up to 11 m tall, but usually about 1–1.5 m in the alpine zone. Bark dark grey to blackish. Branches slender, ringed with the scars of fallen leaves.
- *Leaves*: Crowded towards the tips of the branchlets, 7.5–25 cm long by 3–7 mm wide, tapering to a drawn-out, pointed tip.
- *Flowers*: In clusters on short branchlets just below the tufts of leaves; white, tubular, 7–8 mm long.

Distribution & Habitat
- North, South and Stewart Islands in coastal to low alpine regions from about East Cape southwards. Sea level to 1200 metres.
- Quite widespread in most mountain regions, but much commoner in the higher rainfall regions of the South and Stewart Islands. Occurs in open forests and near the tree-line, and is often an important part of subalpine scrub and, sometimes, in mixed snow tussock-scrub.

Notes
- *Identification*: A very variable species, particularly when growing above the tree-line. In some areas the leaves are very narrow and seldom more than 3 mm wide, while in others they may be up to 5 mm or more wide. Forest plants, particularly in southern areas, frequently have wider leaves.
- *Flowering*: Between November and January.
- *Name*: In the North Island its Maori name is inanga, and in the South Island its dialectal form is inaka.

◀ *Dracophyllum longifolium*
Dunedin Botanic Gardens, November 1995

▲ *Coprosma cheesemanii*
Rainbow Ski Area, January 1996

24 *Coprosma cheesemanii*

Family RUBIACEAE

Description
- A prostrate to sprawling or semi-erect shrub to 50 cm or more tall. Branchlets slender, usually numerous.
- *Leaves*: On very short stalks, narrow, about 8–11 mm by 1–2 mm, pointed, deep green to olive green in colour.
- *Flowers*: Small, greenish, and produced singly at the tips of short leafy branchlets.
- *Fruits*: 6–7 mm in diameter, orange to scarlet or red.

Distribution & Habitat
- North, South and Stewart Islands from Mt Hikurangi southwards, 400–1500 metres.
- Occurs in montane to subalpine areas, often in permanently damp areas in tussock grasslands, herbfields or open subalpine scrub.

Notes
- *Flowers*: As is typical with all *Coprosma* species the male and female flowers occur on separate plants.
- *Fruiting*: Can also be erratic, often depending on whether the season has been suitable for the pollination of the female flowers. Plants are usually in fruit between February and April.

▲ *Celmisia semicordata*
Jack's Pass, December 1995

▼ *Celmisia semicordata* ssp. *aurigens*
Queen's Park, Invercargill, November 1991

25 *Celmisia semicordata*
Mountain Daisy / Cotton Plant / Tikumu
Family ASTERACEAE

Description
- A robust herbaceous plant forming large rosettes, or bold clumps of silvery-green, sword-shaped leaves.
- *Leaves*: 30–60 cm long by 4–10 cm wide, stiff and of a leathery texture with the upper surface silver or silvery-green, the undersurface white.
- *Flowers*: The stout flower stem is up to 50 cm long and the flower heads are 4–10 cm in diameter.

In addition to the typical form, two **subspecies** exist:
- ssp. *stricta*: Has narrower and more rigid leaves with heavily rolled margins; their upper surfaces are also very silvery.
- ssp. *aurigens*: The upper surfaces of the leaves are varying shades of gold.

Distribution & Habitat
- South Island, usually in subalpine to low-alpine regions. 600–1400 metres.
- Common in snow-tussock grassland, herbfields and open subalpine scrub.
- *Typical form*: Occurs throughout the higher rainfall regions from Nelson to South Canterbury and northern Fiordland. Sporadic in Nelson and northern Westland.
- ssp. *stricta*: is restricted to the mountains of western Otago and adjacent Southland to the Takitimu Range.
- ssp. *aurigens*: Occurs in Central and eastern Otago, with outlying populations in south-eastern Otago, and in Southland on the Garvie Mountains.

Notes
- *Identification*: In all of its forms this is an outstanding plant which should be immediately recognised, even if not in flower.
- *Flowering*: Between December and January, depending on altitude.

26 *Celmisia sessiliflora*

Family ASTERACEAE

Description
- Sub-shrubby and usually forming dense, low mats or cushions up to about 10 cm high, varying from 20 cm to 1 metre or more in diameter.
- *Leaves*: Clustered around tips of branchlets to form dense rosettes; rigid, 1–2 cm long, 1–2 mm wide, silvery-white to greenish-white.
- *Flowers*: Heads 1.5–3 cm in diameter, sunk amongst the tips of rosettes.

Distribution & Habitat
- South and Stewart Islands in subalpine to high-alpine regions throughout. 700–1800 metres.
- In a variety of situations from low tussock grasslands, short snow grass, herbfields, fellfields and cushion bogs.

Notes
- *Identification*: Distinct and easily recognised as it often forms extensive patches in open situations such as dry ridges or permanently damp places.
- *Flowering*: During December to January or later.

▼ *Celmisia sessiliflora*
Altimarloch, February 1996

27 *Leptinella atrata*

Family ASTERACEAE

Description
- Rather fleshy herb which forms small clumps or scattered tufts of foliage. The creeping rhizome, quite deeply buried in the shingle scree, puts forth branching stems, each terminating with a tuft of leaves when above the scree surface.
- *Leaves*: 1.5–3 cm long and 7–20 mm wide, finely divided, greyish, margins tinted reddish.
- *Flowers*: Button-like heads are produced singly on short stalks which arise from the tufts of leaves. Heads in bud are dark maroon to almost black; as each flower opens, heads become ringed with bright yellow stamens.

Distribution & Habitat
- South Island: drier mountain areas east of main divide from Marlborough to north Otago. 1000–2000 metres.
- Confined to shingle screes, mainly on the more stable slopes but also on some less stable.

Notes
- *Identification*: One of a unique group of some 21 species generally confined to the harsh environment of the high shingle screes. Non-flowering plants are often difficult to see.
- *Flowering*: Between December and January.

▼ *Leptinella atrata*, Mt Olympus, January 1977

▲ *Raoulia bryoides*
Rainbow Ski Area, January 1996

28 *Raoulia bryoides*
Vegetable Sheep

Family ASTERACEAE

Description
- A shrub of dense and very compact habit which forms irregularly shaped, cushion-like masses 20–30 cm or more in diameter and 10–15 cm thick. The branchlets are so tightly packed together that only their growing tips are visible.
- *Leaves*: Densely clustered together around the tips of the branchlets. Covered on both surfaces with a dense, brush-like mass of white hairs.
- *Flowers*: Flower heads are about 3 mm in diameter and are sunk amongst the foliage at the tips of the branchlets.

Distribution & Habitat
- South Island in low- to high-alpine regions from eastern Nelson and Marlborough to Central Otago. 1200–1800 metres.
- Usually on rock faces and rock outcrops, but also occurs on shattered, but stable, rock jutting out of screes.

Notes
- *Identification*: This, along with several related species, is one of the most extraordinary of our native shrubs and possibly in the world. From a distance it has a white or greyish appearance, but upon close inspection it is most fascinating and rather beautiful. Only the tips of the leaves at the ends of the branchlets are visible, and through the white hairs the bluish-green colour of the leaves creates a rather beautiful effect. With the leaves and branchlet tips all pressed into a hard mass the plant has an almost perfect protection against the elements of its harsh environment. Inside the cushion, protected from sun and wind, the old leaves form a rotting, sponge-like mass which holds water and ensures that the plant does not suffer from lack of moisture. Its main roots are usually deeply anchored into a rock crevice.

29 *Raoulia subsericea*

Family ASTERACEAE

Description
- Usually forms compact mats 30–40 cm or more across. Stems much-branched and rooting into ground.
- *Leaves*: 3–6 mm long, 1–1.5 mm broad, wide-spreading and forming rosette-like tufts at branchlet tips. Upper surfaces deep to medium green; undersurfaces clothed with a thin tomentum of silvery or golden hairs.
- *Flowers*: Heads prominent, up to 1 cm in diameter, produced at branchlet tips. Each head is surrounded by white, papery bracts which resemble petals. The actual florets are a creamy colour.

Distribution & Habitat
- South Island in montane to subalpine regions throughout, except for the wettest regions. 400–1500 metres.
- Common in grasslands and open places, especially in drier hill and mountain country.

Notes
- *Scent*: Flowers distinctly scented, particularly in warm sunny conditions.
- *Identification*: Even when dead, flowers remain a conspicuous feature.
- *Flowering*: Between December and March.

▼ *Raoulia subsericea*, Queen's Park, Invercargill, November 1991

30 *Leucogenes grandiceps*
South Island Edelweiss

Family ASTERACEAE

Description
- A sprawling or semi-erect herb, with a woody rootstock and leafy stems.
- *Leaves*: Closely placed along the stems, spreading or somewhat curving outwards. Leaf 5–10 mm by 2–4 mm, both surfaces clad with silvery or white, closely appressed hairs.
- *Flowers*: Heads in clusters of five to fifteen; clusters 5–15 mm in diameter and surrounded by up to fifteen densely woolly bracts (modified leaves).

Distribution & Habitat
- South and Stewart Islands in low to high alpine regions throughout. 800–1900 metres.
- Most common on rock and stable debris in fellfields.

Notes
- *Related species*: A charming and distinctive plant which superficially resembles the Northern Hemisphere edelweiss (*Leontopodium*), but there the resemblance ends. It is more closely related to the native *Raoulia*.
- *Flowering*: Between November and March.

▼ ***Leucogenes grandiceps***
Otira Valley, January 1996

▲ *Helichrysum bellidioides*
Dunedin Botanic Gardens, November 1994

31 *Helichrysum bellidioides* Everlasting Daisy

Family ASTERACEAE

Description
- A much-branched prostrate or trailing subshrub with slender stems; sometimes forming quite large patches.
- *Leaves*: Closely placed along the stems, spreading, 5–8 mm by 3–5 mm, more or less oval, upper surface usually green, whitish beneath.
- *Flowers*: Produced singly, on short stalks, from the tips of the branchlets. Flower heads 1.5–3 cm in diameter, including the surrounding white, papery, petal-like bracts (modified leaves).

Distribution & Habitat
- North, South, Stewart and Chatham Islands, from East Cape and Mt Taranaki southwards. Sea level to 1600 metres. (Also found in the sub-antarctic: Auckland, Campbell and Antipodes Islands.)
- Occurs in a wide variety of habitats in scrub, tussock grasslands, herbfields, stony places, road banks and rocky outcrops.

Notes
- *Identification*: An easily recognised plant which is fairly common. The flower heads with their surrounding white, papery, petal-like bracts immediately distinguish it. It varies considerably, both in size and other characters: some forms are quite diminutive while others are of a more normal size or of the largest dimensions. Similarly, while the upper surfaces of the leaves are usually green, forms occur on which they are silvery or greyish.
- *Flowering*: Between October and February, according to district. The old flower heads are long-lasting and often give the impression that the plant is still in flower.

▲ *Brachyglottis bellidioides*
Rainbow Ski Area, January 1996

Brachyglottis bellidioides

Family ASTERACEAE

Description
- A small, tufted herb with a rosette of leaves which are often more or less flattened onto the ground.
- *Leaves*: Leaf stalks (petioles) short and hairy, about 1–2.5 cm long. Leaf blades 1–5 cm long, oblong to more or less rounded, apex blunt or slightly pointed, base rounded or narrowing to the petiole; upper surface medium to deep green, clad with short stiff hairs.
- *Flowers*: Flower stems up to 30 cm tall, sometimes branched but not infrequently bearing only one flower. Flower heads 2–3 cm in diameter, yellow.

Distribution & Habitat
- South and Stewart Islands in montane to high alpine regions, almost throughout. 300–1800 metres.
- Occurs in tussock grasslands, snow tussock-herbfield, open shrublands, herbfields and rock bluffs.

Notes
- *Identification*: Often a rather small plant which is frequently tucked in amongst other larger plants. In drier grasslands it more commonly occurs in moist and sheltered sites. When not in flower it is easily overlooked. It is quite a variable species.
- *Flowering*: Between October and March, depending on locality and altitude.
- *Related species*: *Brachyglottis lagopus* is of somewhat similar appearance and is often found in similar situations. It is distinguished by the dense, woolly covering on the undersurfaces of the leaves. Forms of *B. lagopus* occur with leaves up to 20 cm or more long, but generally not in alpine regions. It is found in both the North and South Islands from the Taupo region and the Ruahine Range southwards to South Canterbury and northern Otago, from near sea level to 1500 metres. Common in open shrublands, tussock grasslands and snow tussock-herbfields, especially in rocky sites. Flowering usually occurs between November and February.

▲ *Brachyglottis bidwillii* var. *viridis*
Iron Lake, January 1996

33 *Brachyglottis bidwillii*

Family ASTERACEAE

Description
- A small, compactly-branched shrub from 30 cm to 1 metre tall. Branches stout; branchlets clad with a covering of white to buff felted hairs.
- *Leaves*: 2–2.5 cm long by 1–1.6 cm wide, rounded-oblong or having its widest part above the middle, thick and leathery. Upper surface deep green and shining, undersurface clad with a covering of white to buff, felted hairs.
- *Flowers*: Flower heads lacking petal-like ray florets, 7–15 mm in diameter, produced in tight clusters, up to 5 cm across, on branched flower stems.

Two **forms** are recognised:
- *Typical form*.
- var. *viridis*: A taller shrub growing up to about 1.5 metres, usually more slenderly branched, and has larger and slightly less leathery leaves. The clusters of flower heads are larger and up to 6.5 cm across.

Distribution & Habitat
- North and South Islands in subalpine to high alpine regions. 800–1700 metres.
- *Typical form*: Restricted to the North Island where it is common on the mountains from East Cape and Taupo to Cook Strait. On the volcanoes of the central North Island the typical form also extends into the fellfield.
- var. *viridis*: South Island from Nelson and Marlborough to about the Rakaia Valley. Occurs in low subalpine scrub, snow tussock-herbfield and on rock outcrops.

Notes
- *Identification*: Usually an easily recognised shrub; its compact, branching habit and small, thick, dark green leaves with the white or buff, felted covering on their undersurfaces serving to distinguish it. In scrub and similar more sheltered situations it will grow to its maximum size, but at higher altitudes and in more exposed situations it is usually more compact. Plants growing on exposed rock outcrops are often dwarfed to 20 or 30 cm.
- *Flowering*: Between January and March, according to altitude.

▲ *Cassinia vauvilliersii*
Otira Valley, January 1996

34 *Cassinia vauvilliersii*
Mountain Cottonwood / Tauhinu

Family ASTERACEAE

Description
- A somewhat aromatic shrub varying from less than 1 metre tall up to about 3 metres. Bark of main stems and branches peeling off in small flakes.
- *Leaves*: Numerous, 3–12 mm long by 2–3 mm wide, leathery, upper surface dark green, undersurface with a dull, yellowish or yellowish-brown covering of tightly felted hairs.
- *Flowers*: Flower heads white or creamy, 3–4 mm in diameter, in rather tight clusters of ten to twenty heads at the tips of the branchlets. Each head has several small, white-tipped, petal-like bracts (modified leaves).

Distribution & Habitat
- North, South and Stewart Islands in lowland to low alpine regions from East Cape and Taupo southwards. Sea level to 1500 metres.
- Occurs in open forest and scrub, subalpine scrub and mixed snow tussock-scrub.

Notes
- *Identification*: A rather variable species with several varieties having been named. In some areas it is quite common and may form a conspicuous component of one of the plant associations. Burning and grazing of the vegetation favours its occurrence and its presence often indicates the past treatment of the land.
- *Flowering*: Between November and February. The spent flower heads remain on the bush for quite some time afterwards.

▲ *Gentiana bellidifolia* var. *australis*
Mt Peel (Nelson), January 1996

35 *Gentiana bellidifolia*

Family GENTIANACEAE

Description
- A tufted perennial herb, 10–15 cm tall, arising from a rather stout, single or branched rootstock.
- *Leaves*: Each rootstock or branch has a tuft or rosette of numerous, thick to almost fleshy, overlapping leaves. Leaf blade 1–1.5 cm long by 5–7 mm wide, bright to deep green and shining above.
- *Flowers*: White, 1.5–1.8 cm long, produced singly or in two- to six-flowered clusters.

Two **forms** are recognised:
- *Typical form*.
- var. *australis*: Differs mainly in its stouter habit of growth and larger flowers. It often forms dense, low patches 6–12 cm across, and the abundantly produced flowers are up to 2.5 cm in diameter.

Distribution & Habitat
- North and South Islands in subalpine to high alpine regions from Mt Hikurangi southwards; widespread but often local. 600–1800 metres.
- Usually occurs in damp grasslands, herbfields and boggy places.
- var. *australis* is confined to high alpine regions of the South Island mainly about and west of the main divide. It occurs mainly in fellfields, on rock faces and on bluffs.

Notes
- *Identification*: Of the 24 or so native species of gentian this is one of the finest as well as one of the most commonly seen. The almost fleshy leaves, with deep green and shining upper surfaces, and the prominent heads of quite large flowers are usually sufficient to identify it.
- *Flowering*: Typical form – between January and March; var. *australis* – during February and March. Depending on the season flowering may continue into April.

▲ *Wahlenbergia laxa*
Rainbow Ski Area, January 1996

36 — *Wahlenbergia albomarginata / laxa*
Bluebell / Harebell

Family CAMPANULACEAE

Description
- A small perennial herb with creeping and branching underground stems, and with the leaves forming small rosettes at tips of each stem. Forms small compact patches of numerous rosettes, or more extensive colonies with the rosettes distantly scattered amongst other low vegetation.
- *Leaves*: Vary in size according to habitat, 5–40 mm long, 1–10 mm wide, widest above the middle, more or less spoon-shaped.
  W. albomarginata: margins usually smooth and whitish.
  W. laxa: margins never whitish but with a few small teeth and/or somewhat undulating.
- *Flowers*: Those of each species similar. Flowering stems 10–25 cm long, each bearing a single flower. Flowers 1.5–3 cm in diameter, white to pale flax blue, often nodding.

Distribution & Habitat
- South Island in lowland to alpine regions. Sea level to 1400 metres.
- *W. albomarginata*: from eastern Marlborough and Canterbury to Central Otago, mainly on river terraces and in tussock country.
- *W. laxa*: In higher rainfall regions from N.W. Nelson, along the main divide, in Westland, western Otago and Fiordland. Variety of habitats from river flats and lake shores to subalpine rocks and ridges.

Notes
- *Identification*: Both species occur widely in their areas of distribution. Easily recognised by the small rosettes of leaves which may be thinly scattered amongst other vegetation, or form dense patches up to 20 cm or so across. Formerly, the two species were confused.
- *Flowers*: Colour varies from almost pure white, to white flushed and veined with blue, or sometimes almost completely blue. Size similarly varies.
- *Flowering*: Quite a long flowering season; can extend from November to February.

37 *Donatia novae-zelandiae*

Family DONATIACEAE

Description
- Low-growing and often forms dense, quite wide-spreading, dark green cushions or hummocks, up to a metre or more across.
- *Leaves*: Upper portions of stems clad with very closely placed leaves, 5–10 mm long, 1–1.5 mm wide.
- *Flowers*: White, star-shaped with five, pointed petals; 8–10 mm in diameter, sunk amongst the leaves.

Distribution & Habitat
- North, South and Stewart Islands in subalpine to low alpine regions from the Tararua Range southwards. Common in high rainfall regions. 760–1520 metres.
- In cushion bogs overlying peat, permanently wet depressions in snow tussock or red tussock grasslands and in herbfields.

Notes
- *Identification*: The hard, deep green cushions are easily recognised. The cushions may grow to many metres across and help to form huge cushion bogs. If at all possible, cushions should not be walked on.
- *Flowering*: Between December and March.

▼ *Donatia novae-zelandiae*, Iron Lake, January 1996

Phyllachne colensoi

Family STYLIDIACEAE

Description
- A moss-like plant which forms hard and compact mats or cushions of medium to deep green, 5–10 cm thick and usually 20–50 cm across but sometimes more extensive.
- *Leaves*: Very closely placed on the branchlets, about 4 mm long and have blunt tips.
- *Flowers*: White, 4–6 mm in diameter, sunk amongst the leaves.

Distribution & Habitat
- North, South and Stewart Islands in low to high alpine regions from Mt Hikurangi and the volcanic mountains southwards. 900–1900 metres.
- Usually a cushion plant in herbfields, rocky places, herbmoor and open snow tussock-herbfield.

Notes
- *Identification*: Easily recognised and distinguished from *Donatia novae-zelandiae* by the cushions usually being a lighter, often more yellowish-green and never the deep green of the former species. It can also be distinguished by the flowers sometimes having more than five petals which are rounded at the tips and not pointed.
- *Flowering*: Between November and February.

▼ *Phyllachne colensoi*, Rainbow Ski Area, January 1996

▲ *Pratia angulata*
Dunedin Botanic Gardens, December 1990

▼ *Pratia angulata* fruit,
Queen's Park, Invercargill, April 1992

39 *Pratia angulata*
Panakenake

Family LOBELIACEAE

Description
- A wide-spreading, creeping and rooting herb which often forms large or extensive, matted patches. Stems slender and much interlaced.
- *Leaves*: Variable in size, 4–12 mm by 3–13 mm, more or less rounded with a few rather coarse teeth around the margins.
- *Flowers*: White, 7–20 mm long, produced singly on slender stalks from 2–6 cm long.
- *Fruits*: Magenta or purplish-red berries are rounded or oval and 7–12 mm in diameter.

Distribution & Habitat
- North, South and Stewart Islands in lowland to low alpine regions throughout. Sea level to 1300 metres.
- Common in damp situations in subalpine grasslands, banks, streamsides, open forest and herbfields.

Notes
- *Identification*: Grows in a wide variety of situations, and easily recognised because of the way the flower is split along one side so that the five petals (corolla lobes) are unevenly spaced around it. Not infrequently, ripe fruit can be seen on plants at the same time that they are flowering.
- *Flowering*: Between October and April.
- *Fruiting*: January to July.

▲ *Euphrasia revoluta*
Rainbow Ski Area, January 1996

40 *Euphrasia revoluta*
Eyebright

Family SCROPHULARIACEAE

Description
- A low, tufted perennial herb up to 5 cm tall. Rootstock somewhat woody with several slender, partly-trailing, stems which turn upright near their tips.
- *Leaves*: Narrow, attached to the stems without stalks, 2–10 mm by 1–5 mm, widening towards the tip with one pair of small, sharp teeth just below the large triangular or rounded tip.
- *Flowers*: One to few at the tips of the stems, 1–1.5 cm long, white with yellow in the throat.

Distribution & Habitat
- North and South Islands from the Ruahine and Tararua Ranges southwards, in low to high alpine regions. 900–1700 metres. (Quite widespread in the South Island, but is sometimes inclined to be rather local.)
- Occupies a rather wide range of habitats and may be plentiful in boggy and open places, tussock grasslands and herbfields.

Notes
- *Flowering*: Between December and March.
- *Other*: The species of eyebright are semi-parasitic on the roots of grasses and other plants, although whether any or all of the New Zealand species conform to that pattern is not really known.

◀ *Ourisia macrocarpa*
Queen's Park,
Invercargill,
December 1990

▼ *Ourisa macrocarpa*
ssp. *calycina*
Otira Valley,
January 1996

41 *Ourisia macrocarpa*

Family SCROPHULARIACEAE

Description
- A stout, perennial herb similar to *O. macrophylla*, but usually stouter and more robust.
- *Leaves*: Generally larger than those of *O. macrophylla*, more leathery, and always smooth and hairless. Another distinguishing character is the purple colouring which often occurs on the undersurfaces of the leaves, the leaf stalks and the flower stems.
- *Flowers*: Flower stems are 25–50 cm tall and the flowers are larger, being 2–3 cm in diameter.

Two **subspecies** are recognised:
- ssp. *macrocarpa* which is the typical form and has broad, rounded leaves that are somewhat heart-shaped at their bases.
- ssp. *calycina* is distinguished from the former by its much larger size, the leaves being up to 15 cm long with their bases gradually tapering to the stalk. The flowers are larger and up to 4 cm in diameter.

Distribution & Habitat
- South Island in subalpine to low alpine regions, in high rainfall areas about and west of the main divide. 760–1370 metres.
- *Typical form*: confined to the Fiordland region and occurs in open subalpine scrub, mixed snow tussock-shrubland, snow tussock-herbfield, on rock bluffs and occasionally beside streams at lower levels.
- ssp. *calycina* ranges from northwest Nelson to about central Westland, and occurs in open subalpine shrubland, mixed snow tussock-shrubland, snow tussock-herbfield, on rock bluffs and along streamsides.

Notes
- *Habitat*: In some areas pressure from browsing animals has caused these plants to be restricted to rock bluffs and similar inaccessible places, but where the animals have been brought under control they again become more prominent in other habitats.
- *Flowering*: Between October and February.

◀ *Ourisia macrophylla*
▼ ssp. *robusta*
Queen's Park,
Invercargill,
November 1990

Ourisia macrophylla ssp. *robusta*

Family SCROPHULARIACEAE

Description
- A medium to rather large perennial, leafy herb forming clumps or patches of variable size. Stems stout, prostrate and rooting into the ground. Each branch terminating in a tuft of leaves.
- *Leaves*: More or less erect; leaf blade up to 10 cm long, width usually narrower, upper surface medium green, undersurface paler or sometimes purplish, with few to more numerous fine hairs. Margins with small, blunt toothing.
- *Flowers*: Stems up to 50 cm tall. Flowers in superimposed whorls of four to numerous white flowers, their outsides flushed with purplish-red.

Two **forms** are recognised:
- *Typical form* (*O. macrophylla*): is distinguished by having rounder leaves from 4–10 cm in diameter, a less robust habit and the flowers not being coloured on their outsides.

Distribution & Habitat
- North Island in montane areas from Taupo, the Volcanic Plateau and western Hawke's Bay to the upper Manawatu and the Puketoi Range. The typical form is confined to Mt Taranaki where it is montane to low alpine. Both usually occur within the range of 600–1500 metres.
- *O. macrophylla*: Grows mainly along streamsides and on damp banks.
- ssp. *robusta*: Abundant above the tree-line.

Notes
- *Flowering*: Any time between October and January, although November and December is probably more usual.

▲ *Parahebe lyallii*
Otira Valley, January 1996

43 *Parahebe lyallii*

Family SCROPHULARIACEAE

Description
- A small subshrub with a creeping habit and with slender branchlets which root into the ground.
- *Leaves*: Small, 5–10 mm by 4–8 mm; have up to three pairs of notches or teeth on their margins.
- *Flowers*: Flowering stems are 3–8 cm long and have numerous saucer-shaped flowers about 1 cm in diameter. Their colour varies from white to pinkish with yellowish centres from which pinkish or purple lines radiate outwards.

Distribution & Habitat
- South Island where it is widespread in lowland to low alpine regions. Ascends to 1300 metres.
- Usually grows along streamsides, on stable riverbeds, old moraine, rock bluffs, in snow tussock-shrub herbfield and in similar moist sites.

Notes
- *Identification*: Generally an easily recognised little subshrub which nearly always favours moist situations. Quite often the flower buds are pinkish.
- *Flowering*: Usually occurs over quite a long period between November and March.

▲ *Hebe macrantha*
Queen's Park, Invercargill, November 1991

▼ *Hebe macrantha*
Dunedin Botanic Gardens, December 1995

44 Hebe macrantha

Family SCROPHULARIACEAE

Description
- A short, sparingly branched and rather straggling shrub, usually from 30–60 cm tall. Branches usually erect or sometimes inclined to be spreading.
- *Leaves*: 1.2–2.5 cm by 7–12 mm, thick and leathery; upper surface pale- to yellowish-green, shining, undersurface paler; margins bluntly toothed.
- *Flowers*: White, about 2.5 cm in diameter, tightly clustered near the tips of the branchlets so that the growing tip is concealed.

Two **varieties** are recognised:
- *Typical form*: Has larger leaves measuring 1.5–2.5 cm by 5–10 mm.
- var. *brachyphylla*: Has smaller and more rounded leaves measuring 1.3–1.5 cm by 9–10 mm.

Distribution & Habitat
- South Island, mainly in the wetter regions, in subalpine to high alpine areas. 760–1500 metres.
- *Typical form*: Occurs in the central Southern Alps.
- var. *brachyphylla*: Occurs from Nelson and Marlborough to North Canterbury.
- Both varieties are found on alpine, grassy slopes and in short subalpine scrub on steep, rocky places.

Notes
- *Identification*: This is the largest flowered of the *Hebe* species and as such is easily recognised. Even when not in flower the foliage and its general habit of growth are quite distinct.
- *Flowering*: Between November and March.

▲ *Astelia nervosa*
Jack's Pass, December 1995

▼ *Astelia nervosa*, male flowers,
Queen's Park, Invercargill, October 1991

45 *Astelia nervosa*

Family ASTELIACEAE

Description
- A tufted herbaceous plant which usually forms strong clumps up to 80 cm tall.
- *Leaves*: Numerous, curving or arching, from 50 cm–1.5 m long and 2–4 cm wide. They are quite tough and have a strong and prominent vein running either side of the midrib. The upper surface is a light to medium green and usually has a semi-transparent, silvery covering over it. The undersurface is silvery-white or buff coloured.
- *Flowers*: The branched flower heads are usually more or less concealed amongst the leaves, but when the fruits ripen in late summer or autumn they are quite prominent.
- *Fruits*: 8–15 mm long, various shades of orange, becoming orange-scarlet to almost red towards their tips.

Distribution & Habitat
- North, South and Stewart Islands in lowland to low alpine regions from Mt Hikurangi to Mt Taranaki and Taupo southwards. Ascends to 1500 metres, and in Southland descends to near sea level.
- Occurs in damp ares of mixed tussock grassland-scrub association, tussock grasslands and herbfields, and sometimes in open beech forest.

Notes
- *Identification*: A common and widespread species in most areas of hill and mountain country. There is often considerable variation of leaf colour, and clumps may be rather small or large, according to altitude. In some regions it is a very conspicuous plant, its large silvery clumps standing out from amongst other vegetation. It will form quite large colonies.
- *Flowering*: November and December.
- *Fruits*: Ripen from about February or March onwards.

◀ *Bulbinella hookeri*
Jack's Pass,
December 1995

▼ *Bulbinella angustifolia*
Old Man Range,
December 1995

46 *Bulbinella hookeri*
Maori Onion / Golden Star Lily
Family ASPHODELACEAE

Description
- Generally a rather robust, tufted herb from 30–60 cm tall at flowering.
- *Leaves*: Numerous, mostly erect but may also be somewhat drooping; up to 60 cm by 3 cm, bright green and shining or grey-green and dull.
- *Flowers*: Flowering stems rather stout, usually overtopping the leaves. Flowering portion up to 40 cm by 5 cm. Flowers bright yellow, to 1.5 cm in diameter, numerous and crowded. After flowering the shrivelled remains of the flowers remain attached to the seed capsules.

Distribution & Habitat
- North Island from Mt Taranaki and the Huirau Range to the central Volcanic Plateau and the north-western Ruahine Range, and in the South Island from Nelson and Marlborough to just north of the Hurunui River in North Canterbury. 150–1520 metres.
- Often very common in moist sites in tussock grasslands and herbfields, especially in seepages and other damp places.

Notes
- *Identification*: *Bulbinella* is deciduous, dying down to a tuberous rootstock in the winter and putting forth new growth each spring. The shrivelled and blackened leaf remains often indicate its presence during the winter. In good conditions it will sometimes grow to a metre tall at flowering, and to see a hillside covered with such plants can be a glorious sight.
- *Habitat*: It is unpalatable to browsing animals and its tuberous rootstock is protected from fire so that it is particularly abundant in areas that were formerly grazed or burnt.
- *Name*: Because of its tenuous relationship with the onion it was inappropriately given the common name of Maori onion.
- *Related species*: East of the main divide and south of the Hurunui River the smaller *B. angustifolia* occurs. It is found in montane to low alpine areas as far as Southland.

47 *Chionochloa rubra*
Red Tussock

Family POACEAE

Description

- A usually tall, tufted grass forming large tussocks 80 cm to 1 m or more tall with crowded, narrow leaves.
- *Leaves*: 50–100 cm long and about 2–2.5 mm in diameter, strongly rolled to give a rush-like appearance, tawny coloured to yellowish-brown or coppery-orange, their tips often whitened and the basal areas usually green. Sheath at the base of leaf coloured medium to dark brown. Upon drying the old leaf sheaths either remain entire or fracture into short segments.
- *Flowers*: Flowering stem up to 1.5 metres tall, loosely branched near the top with numerous little spikelets on the branchlets.

Three **subspecies** are recognised: ssp. *rubra*; ssp. *occulta*; and ssp. *cuprea*.

Distribution & Habitat

- North, South and Stewart Islands where it is may be widespread in low alpine regions. 900–1500 metres.
- Often in poorly drained, often peaty low areas, terraces or rolling slopes, generally, but not always, below the tree-line.
- ssp. *rubra* occurs southwards from the North Island Volcanic Plateau and Mt Taranaki to Marlborough and North Canterbury.
- ssp. *occulta* is confined to Nelson and Westland.
- ssp. *cuprea* occurs from mid Canterbury southwards to Stewart Island and westwards to Fiordland.

Notes

- *Identification*: Distinguished by its size, long and narrow, rush-like leaves (caused by the margins being strongly rolled downwards and inwards), and their usually flowing appearance. A further aid to identification is their colour which varies from a tawny to coppery-orange and sometimes reddish colour. Superficially, the leaf sheaths may appear to be a medium brown, but a close inspection will show them to be dark brown.

  Red tussocks are a distinctive and prominent plant of New Zealand's mountain and hill country.

◀ *Chionochloa rubra* ssp. *rubra*, Upper Wairau River, March 1996

Further Reading

Fisher, M.E., Satchell, E. and Watkins, J.M.
Gardening with New Zealand Plants, Shrubs and Trees.
Collins, Auckland, 1970.

Mark, A.F. and Adams, N.M.
New Zealand Alpine Plants.
A.H. & A.W. Reed, Wellington, 1973.
(Godwit Publishing, Auckland, 1995.)

Metcalf, L.J.
The Cultivation of New Zealand Trees and Shrubs.
Reed Publishing, Auckland, 1991.

Salmon, J.T.
Native New Zealand Flowering Plants.
Reed Publishing, Auckland, 1991.

Index of Common Names

Alpine club-moss	1	Mountain cottonwood	34
Alpine hardfern	2	Mountain lily	6
Bidibidi	12	Mountain totara	3
Biddy bid	12	Native violet	8
Biddy biddy	12	New Zealand angelica	16
Bluebell	36	Panakenake	39
Cotton plant	25	Patotara	22
Edelweiss, South Island	30	Penwiper	7
		Piripiri	12
Everlasting daisy	31	Pygmy pine	4
Eyebright	40	Red tussock	47
Golden star lily	46	Shepherd's lily	6
Grass tree	23	Snowberry	19
Harebell	36	Snow totara	3
Inaka	23	South Island edelweiss	30
Inanga	23	Spaniard	17
Korikori	5	Speargrass	17
Maori onion	46	Tauhinu	34
Mt Cook lily	6	Tikumu	25
Mountain buttercup	6	Vegetable sheep	28
Mountain daisy	25	Willow herb	10

Index of Scientific Names

Acaena caesiiglauca	12	*Gentiana bellidifolia*	35
glabra	12	var. *australis*	35
inermis	12	*Geranium sessiliflorum*	9
microphylla	12	*Geum cockaynei*	
Aciphylla colensoi	17	(*G. parviflorum*)	13
monroi	18	*Gingidia montana*	16
similis	18	*Lepidothamnus laxifolius*	4
Anisotome haastii	15	*Hebe macrantha*	44
Astelia nervosa	45	var. *brachyphylla*	44
Blechnum pennamarina	2	*Helichrysum bellidioides*	31
Brachyglottis bellidioides	32	*Leptinella atrata*	27
bidwillii	33	*Leucogenes grandiceps*	30
bidwillii var. *viridis*	33	*Leucopogon fraseri*	22
lagopus	32	*Lycopodium fastigiatum*	1
Bulbinella angustifolia	46	*scariosum*	1
hookeri	46	*Notothlaspi rosulatum*	7
Carmichaelia monroi	14	*Ourisia macrocarpa*	41
Cassinia vauvilliersii	34	ssp. *calycina*	41
Celmisia semicordata	25	*macrophylla*	42
ssp. *aurigans*	25	ssp. *robusta*	42
ssp. *stricta*	25	*Parahebe lyallii*	43
sessiliflora	26	*Pentachondra pumila*	21
Chionochloa rubra	47	*Phyllachne colensoi*	38
ssp. *cuprea*	47	*Pimelea oreophila*	11
ssp. *occulta*	47	*Podocarpus nivalis*	3
ssp. *rubra*	47	*Pratia angulata*	39
Coprosma cheesemanii	24	*Ranunculus insignis*	5
Cyathodes colensoi	20,21	*lyallii*	6
Donatia novae-zelandiae	37,38	*Raoulia bryoides*	28
Dracophyllum longifolium	23	*subsericea*	29
Epilobium macropus	10	*Viola lyallii*	8
Euphrasia revoluta	40	*Wahlenbergia*	
Gaultheria depressa	19	*albomarginata*	36
var. *novae-zelandiae*	19	*laxa*	36

EPILEPSY EXPLAINED

EPILEPSY EXPLAINED

MARY V. LAIDLAW, S.R.N.
Rehabilitation Adviser, The Epilepsy Centre,
Quarrier's Homes, Bridge of Weir, Scotland

JOHN LAIDLAW, F.R.C.P. (Edin.)
Consultant Physician, The Epilepsy Centre,
Quarrier's Homes, Bridge of Weir, Scotland

CHURCHILL LIVINGSTONE
EDINBURGH LONDON AND NEW YORK 1980

CHURCHILL LIVINGSTONE
Medical Division of the Longman Group Limited

Distributed in the United States of America by
Churchill Livingstone Inc., 19 West 44th Street, New York,
N.Y. 10036, and by associated companies,
branches and representatives throughout
the world.

© Longman Group Limited 1980

All rights reserved. No part of this publication
may be reproduced, stored in a retrieval system,
or transmitted in any form or by any means,
electronic, mechanical, photocopying, recording
or otherwise, without the prior permission of the
publishers, (Churchill Livingstone, Robert Stevenson
House, 1-3 Baxter's Place, Leith Walk,
Edinburgh, EH1 3AF).

First published 1980

ISBN 0 443 01962 2

British Library Cataloguing in Publication Data
Laidlaw, Mary V
  Epilepsy explained.
  1. Epilepsy
  I. Title II. Laidlaw, John
  616.8'53   RC372   79-42769

Printed in Singapore by Huntsmen Offset Printing Pte Ltd

PREFACE

We have written this handbook for those who are concerned directly and personally with epilepsy: for patients, their families, close friends, teachers, and for those with whom they work. To write of 'people with epilepsy' as 'patients' does not imply that they are ill but it is less cumbersome and it avoids the word 'epileptics' which seems to set them apart from other people.

We have tried to explain what happens in a fit so that you need not imagine that it is something mysterious, to be feared. We have told you why doctors carry out certain tests and about the pills which patients need to take. There is some practical advice about how relations and friends can help, how the patient can help himself, and how best he can set about finding and keeping a job. Finally, we have tried to explode various myths which cause much anxiety and unhappiness: that, for instance, epilepsy causes madness or an odd personality, that it is responsible for crime, or that people with epilepsy should not marry or have children.

It is often simpler and more accurate to use the proper medical words. As you may not be familiar with them, they are printed in italics the first time they are mentioned

and they are explained in a glossary at the end of the book.
A little book like this cannot explain everything or solve all your problems but we hope it will help.

M.V.L.
J.L.

Bridge of Weir, Scotland, 1980

CONTENTS

1. Kinds of fits   1
2. The purpose of the doctor's examination
   and tests   10
3. Treatment with tablets   20
4. How relations and friends can help   25
5. Does my epilepsy make me odd? Might
   I go mad?   31
6. How can I help myself?   36
7. What will people think of me?   44
8. Is epilepsy associated with crime?   52
9. Can I marry and have children?   55
10. How can I get and keep a job?   62

Appendices
1. The brain   69
2. Driving licences   75
3. Commonly used antiepileptic drugs   77
4. How to get help   78
5. Glossary   81

DEDICATION

To the residents of Chalfont Centre for Epilepsy, from whose example we learned so much, and to the residents at Quarrier's Epilepsy Centre, whom we hope we are helping.

1. KINDS OF FITS

A fit is due to a temporary disturbance in the brain; and there are different types of fit, depending upon which part of the brain is affected. It is easier to understand epilepsy and to get rid of some of the fear and mystery if you have some understanding of the working of the brain. In Appendix 1 there is a simplified explanation of the very complicated brain and of how fits occur when different parts of it are disturbed. You may find it helpful to read this before going on to the description of the different types of fit.

Different types of fit

A patient may have more than one kind of fit, but for any one person the pattern is usually fairly constant, although it may be altered by changes in treatment. If the part of the brain concerned with consciousness is affected at the beginning, the patient will lose consciousness at once and so will have no memory of the attack. This happens in the most severe (*grand mal**) and the least severe (*petit mal*) kind of fit.

---

*Medical terms in italic type are explained in the glossary at the end of the book.

Primary grand mal (generalised convulsion)

The whole brain is involved. The patient loses consciousness and falls to the ground. All his muscles contract and he becomes rigid (*tonic phase*). The contraction of his chest muscles forces air out of his lungs and may cause a weird and terrifying cry. The bladder contracts and he often wets himself—less commonly he soils himself. He may be pale at first but soon becomes blue because he is not breathing. If you are watching, this tonic phase seems to last for ever and the patient may appear to be dead. Actually it only lasts a few seconds. Then the muscles relax, and contract again violently. The alternate slackening and tightening produces jerking movements or convulsions which often go on for several minutes (the *clonic phase*). With each contraction there may be a grunt and frothing at the mouth as air bubbles through the saliva. This may be tinged with blood, if the tongue or cheek have been bitten. Gradually the convulsions subside. The movements become less violent and the periods of relaxation longer. When they stop altogether, breathing becomes normal and his colour returns. However, the brain is exhausted and the patient remains deeply unconscious, in a *coma*, from which he cannot be roused. Within a few minutes consciousness begins to return, and, exhausted, he usually passes into a deep sleep. If he is disturbed at this stage before his brain is working properly, he is likely to be confused and irritable. When he does come round he will feel exhausted; his body will be aching after all the unusual movements and he is likely to have a headache. It is very frightening to watch a grand mal fit but it is important to remember that the patient is unconscious all the time, is quite unaware of what is happening and will have no memory of it.

Petit mal (absence)

Only the part of the brain concerned with consciousness is affected. The patient, who is nearly always a child, loses consciousness suddenly for some 15 to 20 seconds and then recovers immediately. There are no movements apart, perhaps, for a flickering of the eyelids. The child does not fall and he carries on with what he was doing as soon as the attack is over. These absences are such little fits that they often pass unnoticed and occasionally they may account for the child who seems to be day-dreaming at school.

Partial fits

When a part of the brain has been damaged, even if only slightly, a fit may start in the damaged area. The disturbance may then spread to involve the whole brain when consciousness will be lost and there will be a generalised convulsion as has been described already. However, at the beginning of the attack the patient may be aware, fully or partly, of what is going on. There are many different kinds of these *partial fits* depending on the part of the brain where the fit starts. They can be divided into *simple partial fits* when the part involved has a fairly straightforward job to do, and *complex partial fits* when the fit starts where the working of the brain is more complicated and less well understood.

Simple partial fits

These are not very common. If the fit starts in that part controlling movements, there will be a twitching or jerking of a small, but important, part of the body on one side, often the thumb or the angle of the mouth (the left side of the brain controls the movements on the right side of the body and vice versa). The attack commonly spreads to involve movements of the hand, the arm and then the whole of one side. It may stop at this stage, but if it spreads further the whole brain will become involved and then there will be a generalised convulsion.

A neighbouring part of the brain appreciates sensations or feelings. If this is affected, there may be tingling or pins and needles, again often in the thumb or the angle of the mouth. The spread will be the same as for the partial movement attacks.

Complex partial fits

The parts of the brain which lie under the temples are called the *temporal lobes*. These lobes receive the sensations of sound, taste and smell. Also, they are concerned in a much more complicated way with movement and the understanding of sensations. The parts affected in simple partial fits control simple movements or appreciate simple sensations. For example: the movement of clenching the fist, or the sensation that your foot is cold. On the other hand, the temporal lobe organises a series of complex movements such as those necessary to button up your coat or take off your trousers. Parts of the brain closely related to the temporal lobes automatically control the workings of the body—the beating of the heart, breathing, sweating, movements of the gut, and so on. Finally, the temporal lobes have close connections with parts dealing with emotions, memory and consciousness.

Fits are particularly likely to start off in the temporal lobe, if it is damaged. So-called *temporal lobe epilepsy* is, therefore, very common and is quite the most important kind of complex partial fit. It takes a great many different forms. Some attacks are very slight and may be mistaken for a petit mal absence. Others which involve a large part of the temporal lobe and its connections may be prolonged and may spread to involve the whole brain to give a generalised convulsion. Complex attacks vary so much that it is not possible to describe a typical one. There may be the imagination of strange and often unpleasant tastes or smells. If there are movements, they will not be jerky but will seem to be more or less planned, but purposeless. They can last a short time, making the patient fumble with buttons, for example, or they can be prolonged as when a patient walks into a shop, picks up two saucepans and a kettle and walks out again without trying to hide his 'purchases'. It is common for there to be some involvement of the internal control of the body. There may be a change in heart rate, pallor or sweating. The gut may rumble and wind may be passed or the patient may belch. Quite often there is a feeling in the stomach, which is difficult to describe, or a sensation of choking. Memories may be awakened. Sometimes whole scenes unfold like an action replay of some episode from the past. Quite often the patient has an eerie feeling that 'what is happening has all happened before', this is called *déjà vu*. Some attacks are coloured by emotions, which are more often unpleasant, and which range from a general feeling of uneasiness to fear or horror.

Whatever form the attack may take there nearly always is some alteration of consciousness from the beginning. For this reason the patient does not have an ordinary awareness of sensations, they appear distorted and are often difficult to describe. In the same way, anything which he does during an attack is both out of his control and inappropriate. He is not responsible for what he has done, and after the attack he has only the most confused memory of it.

The working of the temporal lobes and their complex connections is largely a mystery to the experts. Complex partial attacks are often very difficult for someone who is watching to understand. They seem rather frightening because the patient appears to be behaving normally, but then, he is not. They may be even more distressing for the patient who may have retained enough consciousness to know that something has happened, but not enough to understand what it was that did happen. When there is some memory left of an attack this is more likely to be an unpleasant one.

Aura

Some people are said to have a warning or *aura* that a fit is coming. The aura is in fact that part of a partial fit which the patient can remember, the part before he loses consciousness. Since primary grand mal and petit mal attacks start with sudden loss of consciousness it follows that there is no aura. Whether or not there is an aura of partial attacks will depend on how far consciousness is altered at the beginning. There is usually an aura of simple partial attacks, for instance movement of the thumb or a tingling sensation, and the patient is often conscious enough to be able to take precautions so that he does not hurt himself when the fit develops. The auras of complex partial attacks are much more varied and include the sensations already described: a smell, a taste, a strange feeling in the stomach, déjà vu, and so on. However, since there is nearly always some alteration of consciousness at the beginning, the patient may well be too confused to protect himself.

Except for petit mal attacks, any type of fit may spread to produce generalised convulsions. Such secondary grand mal is much commoner than the primary kind. Anti-epileptic drugs are much more effective in controlling grand mal and petit mal attacks than partial ones. Once a patient has started treatment he may have very few primary generalised convulsions and his partial attacks will be much less likely to spread to become generalised.

Febrile convulsions

The brains of very small children are sensitive and they may have generalised convulsions when they are stressed by a high fever. Although these *febrile convulsions* are very alarming to the family they are seldom dangerous and most children grow out of them as their brains mature and become more stable. About one in ten children does go on to have fits in later life. This is likely to happen if the convulsion goes on for a long time, as then there may be some damage to the temporal lobes and the patient will be liable to have complex partial fits when he is older. You should always tell your GP if a baby has had a convulsion. If it goes on for more than a minute or two you must get help at once. If your GP is not available, you must take the baby to the nearest hospital. A doctor will be able to stop the convulsions with an injection and so greatly reduce the likelihood of damage being done.

Status (epilepticus)

When a patient has one generalised convulsion after another without recovering consciousness inbetween, he is said to be in *status epilepticus*. Status must be treated at once as it is dangerous. The sooner you are able to take the patient to a doctor who can start treatment the more easily will he be able to control the fits. It is most important not to waste time trying out home remedies. Call your GP or take the patient to hospital on a 999 call. The family need not live in fear that a patient with epilepsy is going to go into status. It is an uncommon condition and there is nearly always an obvious reason. Most often it is due to a sudden reduction in antiepileptic drug treatment. This is why drugs must never be altered without your doctor's instructions and why doctors often arrange for patients to be admitted to hospital if they feel it is necessary to make important changes in treatment.

Why do some people have fits?

Everybody is liable to have fits but only about one person in every 200 actually does. Why is this person unlucky? There are two causes which combine to produce fits.

1. A tendency to have attacks. People are born with brains with widely differing sensitivities. Some have very stable brains and are most unlikely to have fits. At the other extreme there are those whose brains are so sensitive that they may have attacks for no obvious reason. Brain sensitivity also varies throughout life, being high in very small children and rising again at adolescence. As patients get older, their brains become less sensitive and so fits usually get better with increasing age.

2. *Damage to the brain.* If a part of the brain is damaged that part will not work properly and may start off partial fits which may spread to give secondary generalised convulsions. There are many ways in which the brain may be damaged: injuries from traffic accidents, during wartime, or after a particularly difficult birth. Infections may damage the brain itself (*encephalitis*) or the membranes which cover it (*meningitis*). As people get older the vessels carrying blood to the brain become 'furred up' like an old plumbing system and parts may be damaged because

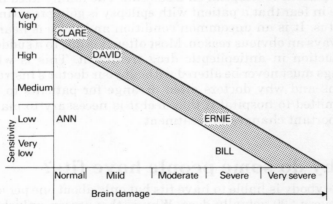

Fig. 1 Diagram to show that fits are due to a combination of a sensitivity of the brain and brain damage. The heavy diagonal line divides those people who have fits (Clare, David, Ernie) from those who do not (Ann, Bill).

they do not get enough blood. Rarely, growths (*tumours*) may cause fits which start in adult life.

Figure 1 shows how the two factors of sensitivity and brain damage combine to cause fits. The people represented in the clear area (Ann and Bill) are the lucky ones who do not have fits. Those in the shaded area (Clare, David and Ernie) are the unlucky ones who do. Ann is well clear. She has low sensitivity and no brain damage. Bill is only just out of the danger zone. He has had severe brain damage but escapes because he was born with very low sensitivity. Clare with no brain damage at all has fits (probably primary grand mal or petit mal) because her sensitivity is so high. David has fits because mild brain damage is combined with high sensitivity. Finally, Ernie's fits occur because severe brain damage is enough to upset even his low sensitivity. The heavy diagonal line divides those who do from those who do not have fits: the unlucky from the lucky. The line is really quite a thin one and it shows the folly of talking of people with epilepsy as 'epileptics' as if they were a race apart.

2. WHAT IS THE PURPOSE OF THE DOCTOR'S EXAMINATION AND TESTS?

You will have seen from the last chapter that fits can take many forms apart from the generalised convulsions which most people think of as 'epilepsy'. It is most important to consult your GP if you or your child has had an attack which might be an epileptic fit. What your GP does next will depend on the age of the patient and on what he finds when he hears the story and carries out his examination. In many cases he will feel that it is necessary to get the help of a consultant who will arrange various further tests. All this can be confusing and worrying if you do not understand what is happening and the way in which the doctors are thinking. They have to decide three things:

1. Is the attack an epileptic fit?
2. What kind of fit is it? This affects treatment.
3. Is there a cause for the fit which can be treated?

Is the attack an epileptic fit?

If the doctor has been called to a child having a febrile convulsion or to a patient whose epilepsy starts with status epilepticus, there will be no doubt that fits have occurred. However, in most cases the doctor will be consulted after the attack is over and the patient is back to his usual self. He will rely entirely on the story that is given to him by someone who saw the attack. It may be difficult to give a good account if you were worried and frightened. Sometimes it helps to make some notes of what you remember before you see the doctor and have to answer his questions. Generalised convulsions seldom present much difficulty although it will be most important to be able to tell the doctor whether they started with sudden loss of consciousness (primary) or whether there was a build up from one of the partial attacks which have been described. Petit mal absences may be missed by the parents for some time. Once they are suspected they are not too difficult to diagnose. Often they occur very frequently and the doctor may see an attack when the child comes to see him. If not, this is a type of fit in which the *electroencephalograph (EEG)*, which will be described later, can be particularly helpful. Simple partial attacks are easy to recognise from a good description. They are not common, a cause for them can usually be found, and often they develop into a generalised convulsion. It is the complex partial seizures which are the most common and important and which cause the greatest difficulty in diagnosis. The EEG is often not very helpful, and, unless he sees an attack for himself, the doctor has to depend on being given a good description. If the attacks are of the kind which develop quickly to a generalised convulsion, there is less difficulty. Often in milder attacks they are not thought to be fits at all and the patient may have had several before he goes to his doctor. Sometimes the fit itself may be very slight and may pass unnoticed but it may be followed by several minutes of confusion during which the patient may act strangely.

The doctor will be particularly interested in two points. Firstly, are the attacks all more or less the same? If so, they are likely to be epileptic. Secondly, do the attacks make sense? Complex partial fits are nonsense: they start for no obvious reason and the way in which the patient behaves during an attack has no meaning and serves no useful purpose. In many attacks the patient will have an aura and, if he can describe it to the doctor, this helps. Some, such as strange unpleasant tastes or smells, are very suggestive. The feeling of déjà vu, although characteristic, may be experienced by people who do not have fits and so is less important. Even at the beginning of an attack there is some alteration of consciousness and so the patient will find it difficult to describe what happened. If his description is vague, muddled and difficult to express, he is likely to have had a fit. If he gives a clear and detailed description it is more likely that he is suffering from some mental upset.

Fits must not be confused with simple faints, which are common, particularly in young women. They are due to not enough blood getting up to the brain. They almost always occur when people are standing, often in hot stuffy conditions. The patient has warning sensations and he loses consciousness gradually. Once he has fallen down he will begin to recover as the blood starts to flow back to the brain.

The doctor has to distinguish between epileptic fits and many other causes of loss of consciousness or abnormal behaviour. Sometimes the blood supply to the brain is affected by occasional disorders of the heart or obstruction to important blood vessels. There is also a group of people whose sudden strange behaviour is due to a mental disturbance. Panic attacks can usually be shown to have been brought on by something which the patient finds frightening. Quite often people who find they cannot cope with life have attacks, which they cannot control, to avoid a situation with which they cannot deal.

What kind of a fit is it?

The doctor will be answering this question at the same time as he is deciding whether the attack is epileptic. It is important for him to know the type of fit and there are various ways in which it may be of practical importance to the patient.

1. Very slight attacks may be petit mal absences or the mildest form of complex partial fits. The treatment is quite different.
2. Some complex partial fits develop so quickly that they may be mistaken for primary generalised convulsions. If this kind of partial fit is treated, the spread may be prevented and the patient may have only a partial fit. This will be a new kind of attack of which he may be vaguely aware and he may be very worried by it if he does not understand what is happening.
3. Some patients may have complex partial fits none of which develop into generalised convulsions. He must not be allowed to think that this is 'not really epilepsy'. It is. There is always the danger that the attacks might become convulsions and it is important that he should take the antiepileptic drugs his doctor gives him to prevent this happening.

Is there a cause for the fits which can be treated?

Fits can always be treated, often with considerable success, although it is uncommon to find a cause for the fits which can simply be removed. It is important to realise that the examinations and tests carried out by GPs and hospital specialists are now very accurate and it is most unlikely that a treatable cause of fits will be overlooked. By visiting many different hospitals and doctors in the hope that they will find out the cause of your fits, treat it and cure you, you are likely to become unnecessarily disappointed and frustrated.

Epilepsy is not a single disease like 'flu or bronchitis. It may be due to many different causes. It follows that doctors deal with each patient differently. Some may need a great many tests. Others may be diagnosed after very few. You must not feel that if 'Mrs McGregor', who had fits, spent three weeks in hospital and had a great many tests, and your young 'Willie' spent only half a day in the hospital and only had an EEG, your doctors have not been looking after Willie properly. Willie is a healthy lad. The doctor could find nothing wrong when he examined him. Blood tests did not show any of the rare conditions which occasionally cause fits. His EEG although it confirmed that he was having fits, did not show that any particular part of the brain was damaged. Mrs McGregor had been out of sorts for some time and had had increasingly severe and persistent headaches. When her GP examined her he found that one side of her body was a bit weak. Her EEG showed that there was something wrong on the other side of the brain (remember that the left side of the brain controls the right side of the body and vice versa). Later in hospital the *EMI scan* (see later) confirmed that there was a suspicious area in the brain. It seemed probable that she was one of the few unfortunate people whose fits were due to a growth (tumour). The further tests were carried out to find out whether she would be suitable to have the tumour removed at operation. There was no need for Willie to have all these complicated tests. All that he needs is to have his fits controlled as well as possible with antiepileptic drugs.

The Electroencephalograph (EEG)

The EEG can be a most useful test in epilepsy and most patients have one done. The equipment is quite expensive and looks a bit complicated and frightening, but the test is quite painless and never does any harm. It is helpful and interesting to understand what it is all about.

Most of the working of the body is controlled by electrical changes in the nerves and muscles. These changes are very small indeed but there are so many millions upon millions of nerves in the brain that the electrical changes can be measured if they are magnified about a million times. The EEG machine does just this. A cap made of thin rubber tubes is put on the patient's head and holds down a number of little pads (*electrodes*) which pick up the electrical changes from different parts of the brain. The electrodes are connected to the EEG machine and the changes are magnified to produce enough power to move delicate ink-writing pens up and down with each change in brain voltage. As the pens write on moving paper, the up-and-down movements show up as waves. When the brain is working normally the voltage changes are small and alter quickly producing the small waves seen in Figure 2. The rubber cup keeps the electrodes firm.

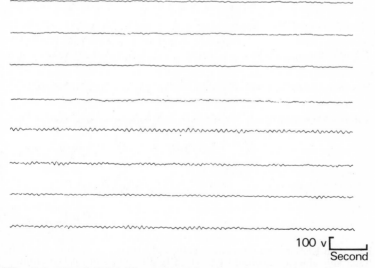

Fig. 2 A normal EEG. The top four lines come from the front of the head, the bottom four from the back.

This is important because if they move much, more electrical disturbance is produced than comes from the brain and large irregular waves result which may be confusing. Small children are often restless and when they move about the electrodes move too. This is why it is sometimes necessary to give a child a simple sedative before the EEG test to keep him quiet.

Understanding the EEG is quite difficult and something for the specialist but two simple examples may help you to appreciate that a fit is not really all that much of a mystery and that it can be seen on paper.

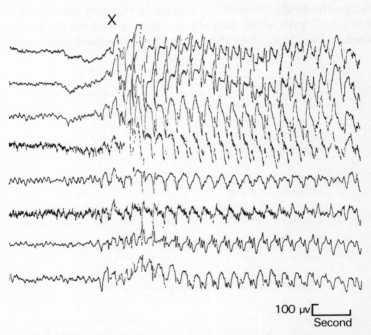

Fig 3 An EEG record showing a petit mal absence in a child. The left hand side shows a normal EEG (slightly different since it is a child). The absence starts at X. The record changes dramatically to large regular slow waves and sharp spiky ones.

1. Figure 3 shows what happens during a petit mal absence. On the left hand side the brain waves are quite normal. Suddenly at the point marked X there is a dramatic change. From all over the brain there are very large slow waves with very spiky waves; this is the absence.
2. Figure 4 shows large slow waves with the same spiky waves coming from some of the lines with quite regular small waves from the others. This means that a part of the brain is not working properly while the rest is all right. This pattern suggests that there is an area of brain damage which is producing the spiky waves and which might cause a simple partial fit.

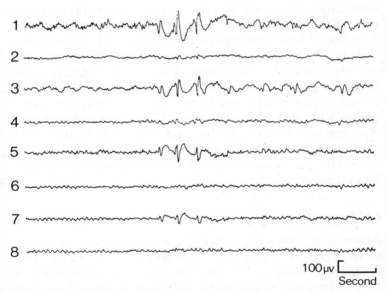

Fig. 4 The odd lines come from the right side of the head, the even ones from the left. The record on the left is normal. That on the right shows both slow and sharp spiky waves suggesting that this side of the brain has been damaged and is likely to be the site of origin of a *simple partial fit*.

Although the EEG may be very useful, it does not give all the answers. A patient with epilepsy may have a normal EEG. This does not mean that he does not have epilepsy but it does make it much less likely that his fits are due to some serious trouble in the brain. Complex partial fits are particularly difficult to diagnose from the EEG whereas for patients having primary generalised convulsions or petit mal absences the EEG is much more helpful.

Towards the end of the EEG recording it is usual to ask the patient to take deep breaths for about three minutes. This produces changes in the blood which make it more likely that abnormalities will be seen. Some patients are sensitive to flickering light. They may have fits when watching a television set which is not working properly. In some EEG departments the patient is asked to look at a very bright flickering light while the record is being made. This is a little unpleasant but it may be useful in bringing out abnormalities suggesting epilepsy. Again, it is quite usual for people to have their fits when they are falling asleep or just waking up. Sometimes it is helpful to carry out an EEG test after the patient has been given a simple sleeping pill. There are other and more complicated EEG examinations but these are only used very occasionally and in the rare cases when it is felt that an operation may be necessary.

Other tests

Fits may be caused by something wrong in some other part of the body which upsets the working of the brain. The doctor will want to do a general examination and take a blood sample. It is unlikely that he will find anything, but if he does, the trouble can often be treated, the fits will stop and the patient will not suffer from epilepsy. For example, fits in infants may be due to lack of calcium or sugar in the blood.

It was explained in Chapter 1 that damage to the brain is an important factor in causing fits. From his examination and the report on the EEG, the doctor may suspect brain damage to be present. It is useful when he is planning treatment to know what this is; very occasionally the damage is due to a tumour which should be removed. An X-ray of the head is very easy to do and it is possible to show up something on this simple test. Within the last few years a very complicated kind of X-ray has been invented which gives a great deal of information about what is going on inside the head. It is usually called the EMI scan since it was invented by the firm of Electrical and Musical Instruments Ltd. All that the patient has to do is to lie quite still on his back for about five minutes while the X-ray machine takes hundreds of serial photographs of the brain from many different angles. These photographs are fed into a computer which comes out with a series of maps of the inside of the head. From these maps the specialist is able to tell whether or not there is something wrong. The EMI scan is an expensive test and is not yet available in all hospitals but it has proved a great blessing to patients since it has replaced other tests which were much more uncomfortable and which carried some slight risk. You must remember that it is only some patients with epilepsy who need an EMI scan and it is not done to find a cure for epilepsy but to make sure that there is nothing serious going on which is causing the fits and which needs to be treated. You must not feel that you are missing a possible cure if your doctor has not felt that an EMI scan was necessary.

In the rare case when the EMI scan shows up something which does need to be treated, the patient will be admitted to hospital and other, more complicated, tests will be done to decide the best treatment. These will not be described here since they only apply to a very few patients and the specialists will explain what they are about.

3. TREATMENT WITH TABLETS

Almost all patients will need to take tablets, often for the rest of their lives. It is very important to follow your doctor's instructions exactly. However, it is not enough to say: 'You must do what your doctor tells you'. Patients are sensible people and they are entitled to know why it is important to do so. If you are going to have to take tablets for many years, it is worth while finding out something about them.

The different drugs

There appear to be a great many different drugs for epilepsy and this is confusing. In fact, however, there are not many important ones. Much of the confusion arises because of trade names and approved names. Drug companies spend a great deal of money developing new and useful drugs. When they find one which has proved successful they give it a trade name and for some time they are the only people who can sell that drug. This monopoly allows them to get back some of their costs of development. If the drug has proved to be really valuable, it is given an approved name. (Unfortunately the approved name is usually more scientific and complicated and doctors who are used to the easier trade name go on prescribing it under that name.) When the monopoly runs out, any other drug firm is allowed to make and sell the same drug and they give it a trade name of their own. Thus the same drug can have several trade names as well as one approved name. The position is made even more complicated because foreign countries have different trade and approved names. For example, Parke-Davis developed a drug which was very effective in treating petit mal absences and called it Zarontin. It was accepted and given the approved name of ethosuximide. Then another firm made the same drug and gave it their own name of Emeside. The more important drugs are given in Appendix 3. Drug firms also have a habit of combining two antiepileptic drugs in one tablet and giving this yet another trade name. Nowadays it is not considered to be a good thing to use these combined tablets and so they are not included in the appendix. However, they do sometimes add to the confusion.

Which drug to choose

This must be a matter for your doctor to decide. If a patient is having very slight attacks, it is important to decide whether these are petit mal absences or very mild partial complex fits. Ethosuximide is the best treatment for petit mal absences but does no good in other types of fit; and conversely most of the drugs used for these other types of fit are no good for petit mal. Apart from this, it is not possible to say that a particular drug is the one which must be used for a particular type of fit.

There is no new wonder drug. All the main ones have been in use for many years, although there are some unusual forms of fits, which were not described in Chapter 1, which may benefit from drugs which have been introduced recently. Different people seem to respond to different drugs, and, if a friend with epilepsy seems to have benefited from Drug Y, when you still have fits on Drug X, this does not mean that your friend has a better doctor than you do. Complex partial fits are particularly difficult to control completely with drugs and it is often better not to try too hard to do so but rather to give just enough tablets to make sure that they do not develop into generalised convulsions.

Over the past few years a great deal has been learnt about what happens to antiepileptic drugs after they have been absorbed into your body. This very complicated matter is for the experts. Briefly; one drug may increase or decrease the effect of another; they may affect or be affected by other medicines which you are taking; and they may influence the internal workings of the body. Therefore, if you are having problems with the control of your fits and if it is some time since you saw a specialist, it is quite sensible to have a talk to your GP and ask him whether he would refer you to hospital for advice.

Side effects

Nowadays, a lot is heard about people addicted to drugs like 'pot' and heroin. There is no danger of becoming addicted to the drugs used to control epilepsy. However, the useful ones are powerful and even in ordinary doses they have some effects other than controlling fits. This is the price that the patient has to pay for having his fits controlled. Usually the price is a small one: a slight slowing down or perhaps difficulty in getting up in the morning. If too large doses or too many different drugs are taken, the side effects can become much more serious and the price much too high. After treatment has been started you may still be having a few fits and these may worry you or prove an embarrassment at work. It is easy to feel that, if you had a few more pills, all would be well, and to press your doctor to increase the dose. If you are having the proper dose already, this is most unwise. Increasing the dose is not likely to make the fits any better and often it may make them even worse. More importantly, there is a real danger of becoming intoxicated with your anti-epileptic drugs in much the same way as a person becomes intoxicated with alcohol. There may be drowsiness and confusion, unsteadiness and difficulty with speech or with fine finger movements. These side effects are fairly obvious but there may be more subtle changes: depression, irritability and in extreme cases mental changes which border on madness.

None of these side effects are likely to occur on ordinary doses of your drugs but they are a very real danger if you press your GP too hard to 'Give me more pills to stop my fits'.

Do not treat yourself. You will have realised the importance of following the doctor's instructions in taking the antiepileptic drugs which are so necessary to control your fits but which have so many possible dangers.

Here are two particular DON'TS:

DON'T increase your tablets because you have had a few extra fits. If you add one or two extra tablets this will do no good. If you make a big increase there is a risk of intoxication.

DON'T stop or reduce your tablets without your doctor's permission. If you do there is a very real danger that you will have severe generalised convulsions or even go into status epilepticus.

Most of the drugs stay in the body for some time and so there is no need to take them more than twice a day. It is usually fairly easy to remember to take them in the morning and the evening, but often more difficult to remember in the middle of the day when you are busy. Therefore, unless your doctor has given you other instructions, it is better to be on a twice daily dose. If you do forget one of your doses, do not panic. You will not go into status and you can take the missed tablets with your next dose.

4. HOW RELATIONS AND FRIENDS CAN HELP

It is always very disturbing to watch anyone having a fit. It is even more frightening when you are watching a member of your family or someone whom you know. Then even a mild attack may be horrifying. The person you know so well behaves in a way which is impossible to understand. He seems to have become someone else. It is not surprising that in the olden days he was thought to have been possessed of the Devil. You should not be ashamed of your fear or horror. It is a very natural feeling. The best way to get over your shock is to do something to help, no matter how little. To watch someone having a major fit is rather like listening to a loud alarm clock going off inside a locked bird-cage. There is nothing that you can do to stop it and so it seems to be ringing for an hour although you know it will stop in a minute. Like the alarm clock, almost all fits stop by themselves and the patient recovers completely. It is a help to know what you should and what you should not do.

Petit mal absences

Nothing needs to be done. If your child has absences, they may be very frequent and some may be so short that you do not notice them. It is important to remember that he may miss out on things you have told him and not blame him for disobedience. Children with absences have normal intelligence but teachers who do not understand may think of them as stupid, lazy or inattentive.

Complex partial fits

These may take very many different forms (see Ch. 1). The patient is not properly conscious and he is not responsible for what he is doing. Unless he is doing something likely to harm himself or someone else, DO NOT interfere with him in any way. If you do, he may react violently because he is confused. Although he may appear to have recovered, he may remain muddled for several minutes. He will be helped by quiet reassurance from someone he knows, like a child who is frightened when he wakes up in a strange place. When he has recovered fully he will usually be aware that he has had an attack but will have no memory of what happened. Intelligent patients are often very worried that they did something dreadful or silly. It is most important for their morale that they should find that everyone has treated the episode quite calmly and not as a catastrophe. Some patients are helped by being told what did happen. There is no need to mention behaviour which might seem to have been too undignified.

Generalised convulsions

As explained in Chapter 1, these may come suddenly (primary) or may develop from one of the other types of fit. They are very alarming, but once they have started there is nothing that you can do to stop them. When the fit has run its full course it will stop quite naturally on its own. When the patient is convulsing there is usually little that you can do, although it may be possible to loosen any tight clothing like a tie or a belt, and take off his glasses if he is wearing them.

During generalised convulsions

DO NOT try to restrain his movement.
DO NOT interfere with him in any way unless he is in obvious danger, like falling into the fire.
DO protect his head from the ground with a cushion or failing that your foot.
DO NOT put your fingers or anything else in his mouth.

After generalised convulsions

When the convulsions have stopped and before he has recovered consciousness it is very important that he should be able to breathe easily, and that saliva does not trickle down the air-passage.
DO remove any false teeth.
DO turn him over on his side into a position as if he were hugging the floor.
DO make sure that his head is also turned to the side. It should be lower than the rest of the body. If the tongue is falling back, it must be pulled forward and the jaw should be pressed upwards. If there is a lot of froth and spit in his mouth, clear this away gently with a cloth or handkerchief.
DO NOT try to give him anything to drink when he is unconscious.
   When you are satisfied that he is breathing normally leave him to recover slowly in his own time. He will be unconscious and so will not feel anything. It may be possible to remove stones, grit or perhaps glass from any part which has been injured, but if you are in any doubt leave this to the doctor. If there is bleeding from an injury, stop this by covering the bleeding area with a clean pad of material and applying simple pressure and later a bandage.
DO remove any evidence of incontinence which will embarass him on recovery.

When the patient comes to, be there to reassure him. When he is recovered fully a cup of tea will probably be welcome. DO NOT tell him how frightening or severe his fit was.

DO NOT call an ambulance or a doctor until you have given the fit a chance to take its natural course.

DO get medical help if:

1. The convulsions go on for more than five minutes.
2. The patient has another generalised convulsion before he has recovered consciousness.
3. He has suffered an injury with which you cannot deal.
4. You cannot stop any bleeding.
5. He does not recover consciousness within 15 minutes of the end of the convulsions particularly if he has hit his head during the attack. Many patients fall asleep after the fit is over. If he is asleep he will make some response when you shake him gently. If he is unconscious he will not.

If a patient has a fit when other people are about, DO all that has to be done as calmly as possible and DO make sure that other people don't do any of the DON'TS.

Although a fit may be frightening, helping the patient who is having an attack is quite straightforward. Helping someone to live with his epilepsy is much more difficult. Some of these problems will be dealt with in Chapters 6 and 7. There are some general ways in which you can help someone in your family:

1. As far as you possibly can, try to take his fits as a matter of course. Unless your doctor has told you to do so, do not keep detailed lists of fits and the times when they happened. Do not let the whole family life revolve around the patient and his fits.
2. If someone in your family has fits, you are bound to be worried that he will come to some harm and will want to protect him. It is not possible here to give detailed

instructions about what restrictions are necessary, since each patient's circumstances are different. However, it is always better to protect too little than too much. Under-protection may run a small risk of physical injury whereas over-protection is sure to cause disturbances of behaviour and possibly of personality later on. It is very important for children not to appear different from their friends and school-mates. If they cannot get on with their age group, they will never learn how to get along with other people and when they grow up they will be isolated. Any personality characteristics will become exaggerated because they are not altered by mixing with other people. It is very important not to make your child feel the odd-one-out at school.

As any child grows up he begins to resent the authority of his parents and usually becomes mildly rebellious as he learns to become independent. The young person with epilepsy may show quite serious behaviour problems if restrictions for his protection are too great. If some restrictions are necessary, it is wise to treat the adolescent as an adult, to discuss the reasons for them with him, and help him to make his own decisions. It can be most dangerous to try to impose restrictions on a sensible adolescent. It is advisable to include the child in any discussion with your GP about the details of necessary precautions. The child will be much happier to accept something which the doctor has said and you will be prepared to accept greater risks if the responsibility is not all yours.

3. The patient in your family (child, husband or wife) is going to need medical care for the rest of his life. It is very important that you should find a GP in whom you and the patient have complete confidence. When you have a good doctor you want to be a good patient. The doctor is human and if you are a bad patient you will not get the best from him. Apart from emergencies, here are some suggestions to help you to become a good patient.

a. Take the antiepileptic drugs exactly as the doctor has told you.
b. Do not just collect a prescription for renewal of tablets. The patient must see the doctor, if only for a few minutes. He can then assess any possible side effects of tablets and you can report any important change in the pattern of fits or anything else unusual. Your GP is busy, however, so avoid unnecessary detail.
c. If you are worried that things are going wrong, take your GP into your confidence and discuss with him whether it might help to get a specialist opinion. But do accept his advice if he says 'No'. Fits can almost always be made better, but seldom can they be cured. Shopping around for a cure from one specialist to another does no good. It raises false hopes in the patient and makes the doctor think that you are a difficult patient.

5. DOES MY EPILEPSY MAKE ME ODD? MIGHT I GO MAD?

A hundred years ago, epilepsy was counted as a form of madness and many 'epileptics' were shut up in asylums. The idea still persists in some primitive communities. Later, there was a very widespread feeling that, although patients were not actually mad, there was something different about them. This fear of possible madness or of some sort of oddness still causes a great deal of distress to patients and their families. It is not very helpful simply to say: 'That is all nonsense; everyone with epilepsy is perfectly normal'. It is much more useful to explain just what connection there may be between epilepsy and disturbances of the mind or peculiarities of behaviour.

Brain damage

As was explained in Chapter 1, brain damage may be an important cause of fits. If severe, the patient may be mentally handicapped, but it is important to emphasise that it is not the fits which have caused the brain damage. It is the brain damage which has caused both the fits and the mental subnormality. Quite a number of patients in hospitals for the subnormal have epilepsy which may be difficult to control, but their main problem is their subnormality. If a patient is of normal intelligence when he starts to have fits, there is no need to fear that he will become subnormal because of his fits.*

The effect of antiepileptic drugs

Drugs in ordinary doses do not cause any important mental upset. They may cause some slowing down and a slight fuzziness in thinking. This may not be important to someone working as a gardener but it would be to an accountant. Nowadays, doctors try to keep the dose down as low as possible to prevent this side effect. It is often better to risk having a few fits and feel mentally alert, than to have no fits and not be able to get on top of your work. This is something which the individual needs to discuss with his doctor. Phenobarbitone, which is a drug used very commonly, often causes children to be very irritable and overactive: you should watch for this and report it to your GP. It is usually possible to change to another drug which is equally useful.

When too large doses of drugs are given, there is a very real danger of quite serious mental upsets. The patient may become very slowed down and drowsy so that he cannot think straight. He may be so depressed that he wants to harm himself. He may imagine that everyone is against him and that he is being persecuted. He may see things which are not there or, less commonly, imagine that voices are talking to him and telling him what to do. His behaviour sometimes becomes quite unreasonable, so that his family cannot cope with him. Any of these things must be reported to your doctor at once. If the dose of drugs is reduced, he will get better very quickly.

Head injury

Very occasionally it proves impossible to reduce fits, even with the most sensible use of drugs and a patient may

*Very occasionally, untreated and severe *status epilepticus* may cause further brain damage.

continue to have frequent attacks. If these are of a type in which he is liable to fall and hit his head, he will suffer a large number of slight injuries. These build up—just as they used to do when professional boxing was less well supervised. The patient will become 'punch drunk', with disturbances of behaviour and loss of his senses. Although this is very uncommon, exceptionally it will be sensible to persuade him to wear some sort of head protection. Nowadays, this can be designed so as not to be too conspicuous.

In connection with the fits

Generalised convulsions

After one of these attacks the patient will be muddled for some time and often behaves strangely. This is not likely to be confused with mental upset since it is obviously connected with the fit. However, if he is forcibly restrained or interfered with during the confused period, he may react violently. Quite often patients are considered, quite wrongly to be dangerous: they are not; and there would be no trouble if they were left alone to recover. During a grand mal fit there is a great deal of disturbance of the activity of the brain. If a patient has a very severe attack or several attacks, it may even take several days before the brain starts to work properly again. During this time his behaviour is sometimes very peculiar. It is important not to get alarmed by this and to realise that he will recover his senses fully. As far as is possible, leave him alone. He may need a good deal of looking after but it is useless to argue or to try to reason with him.

Complex partial fits

These very common attacks take many forms and involve parts of the brain concerned with emotions. It is common for there to be strange behaviour before, during or after an attack. It is easier to understand this if the fit itself is obvious, but if the physical effects are slight, and pass unnoticed, then what is seen is the apparently 'mad' behaviour. Although the patient is not in any way responsible for what he is doing during the attack, he may be partially conscious and aware of what is going on. Sometimes he experiences fear or horror. His impressions of the outside world are distorted because his brain is not receiving normal messages. There is a real danger that he will react like a cornered animal if he is approached. However dangerous his wild actions may seem to be, they are harmless because they are not consciously directed. If the attack is understood and he is left alone, there will be no trouble.

Very occasionally patients who have this kind of fit may behave strangely or seem odd between their attacks. This does not mean that they (or anyone else with epilepsy) are mad. The trouble is sometimes due to damage in the sensitive parts of the brain involved and sometimes due to very little attacks which pass unnoticed.

Personality

It used to be thought that 'epileptics' had a special and rather awkward kind of personality caused by their fits. This false belief was largely responsible for these patients being considered as a group apart, as different from other people. It is still held by some older psychiatrists and causes a great deal of distress. Fits do not alter personality but there are ways in which epilepsy sometimes causes personality difficulties.

People have widely differing personalities just as they have different heights, weights or intelligence. However, as we grow up and develop we mix with other people, at home, at school and at work. There is a natural tendency to want to be able to get on in our group, and so we control or adjust our peculiarities so that we become more acceptable. Our sharp corners get rubbed off like those of pebbles in the bed of a fast flowing river. Many patients, particularly children find it difficult, because of their fits, to become accepted, and they feel isolated. When they leave school any friends they may have made drift away and they are at risk of becoming more alone. Perhaps they cannot get work. They may not have a girl or boy friend. They cannot drive a car. Isolated in this way, there is a danger that the patient may withdraw into himself and that any peculiarities in his personality become exaggerated. It is most important for parents and teachers to do everything possible to help him to be ordinary, even if this means some risk of physical injury in a fit. A broken leg will mend. A disturbed personality will be with him for the rest of his life.

The problems of living with epilepsy and of the reactions of others to fits are considered in the next two chapters. Sometimes these are very formidable. Those of us with ordinary, average personalities can cope with the ordinary, average problems which we meet in everyday life. Sadly, there are still those with epilepsy with quite ordinary personalities who have to deal with very extraordinary problems in their everyday lives. How would you feel if you were cut out of your favourite sister's wedding in case you had a fit and upset the neighbours; or you were told by fifteen employers to whom you had applied for a job, 'I'll let you know', and they never did; or as a brilliant young accountant, you had a partial attack during your first audit and left the floor of an important client covered with your incontinence. A patient with an exceptionally strong personality might cope with these situations. Surely the ordinary average patient should not be condemned as having an 'epileptic personality' if he becomes resentful, frustrated and feels that the world is against him.

6. HOW CAN I HELP MYSELF?

Understand your fits

Although fits will be reduced with proper drug treatment, most patients will have some for many years. It is valuable that the patient should understand his fits as well as possible. Chapter 1 and Appendix 1 on the working of the brain should help. If there are points which you do not understand or you want to know more about your own fits, you should have a talk with your doctor. Do not be put off by the idea that he is too busy. He will be treating you for years and it is well worth the time spent getting everything straight from the beginning. If his surgery is very busy you may be able to arrange an appointment outside surgery hours, or he may ask the Health Visitor or District Nurse to talk to you. The more you understand about your fits the less frightening or mysterious they will be.

You should find out from your family or your friends just what does happen when you have an attack. Often patients imagine that all sorts of terrible things go on when they are unconscious. Families sometimes think it is kinder not to talk about the fits. This is not sensible since it is always better to know what really happens than to imagine the worst. You can then organise your life so that your fits cause the least amount of disturbance rather than allowing them to dominate and disorganise what you are doing. You must remember that it is your life and they are your fits and it is much better for you to decide what precautions to take than to allow restrictions to be imposed on you.

Complex partial fits

These may be particularly difficult to understand because during part of the attack there is only a partial loss of consciousness. The patient may be aware of thoughts, sensations and memories in a confused sort of way. Because they are due to unusual brain activity these are often distorted and frightening. For example, tastes may be experienced which are not only unpleasant but even obscene. There may be indescribable sensations from within the body. The patient may be aware dimly that he is doing strange things over which he has no control. It is most important that he should understand that all this is due to a temporary upset in a complicated part of the brain. Otherwise he will have a real fear that there is something permanently wrong with him and that he is becoming odd or even mad; and this is not so.

Accept your epilepsy

When first you are told that you have epilepsy, you are quite likely to have a disastrous feeling of shock as if you had been told that you had some dreadful disease. This shock reaction is not at all reasonable but it is quite natural. At first you may try to deny the diagnosis and pretend that: 'I don't really have proper fits', or you may become deeply resentful and want to hit out at someone or something as the cause of the trouble. Similar reactions happen when someone whom you love dies. However, it is most important that you should not allow them to carry on after the time of early shock, that you should accept your epilepsy. If you think about it sensibly, you have no alternative. Your fits will be made better with treatment but you will still have to live with your epilepsy. If you do not learn to do so, the effects on you as a person will be far more disabling than the fits. You and your family need to understand that it is quite unusually difficult to adjust to living with epilepsy. For almost all the time you are perfectly well, healthy and able to look after yourself. For perhaps one hour in every month, when you are having fits, you are completely disabled, unable to control yourself. The only time when you can sort things out with yourself is when you are not having fits and when you feel perfectly all right. It is difficult to explain to yourself that you are better off than someone crippled by painful arthritis or disabled by chronic bronchitis. It may help to remind yourself that epilepsy, if properly treated, is hardly ever dangerous; that fits are not painful, although sometimes you may suffer injuries which are. In what ways can you help yourself to accept your epilepsy?

You must make sure that you do not come to have a 'chip on your shoulder'. You must mix normally with other people. There may be incidents of prejudice against epilepsy and unkindness to patients, but there are probably many more occasions when the patient is unwittingly at fault, when, expecting trouble, he becomes on the defensive, almost aggressive and thus quite unapproachable. He is uncommunicative and prickly because he has failed to accept his epilepsy, sullen and resentful. Who wants to know someone like that? If people with epilepsy cannot accept and understand, how can they expect others to do so?

Special help should be given to children with fits. The school is all too often a tough jungle of innocents in which the strong persecute the weak and gain some satisfaction from their teasing. If a child with fits should be subject to such treatment, his best defence is indifference. It is really no fun to tease someone who is unteasable.

Keep interested and active

Fits are much more likely to occur if you have nothing to do or think about. Employment is considered in Chapter 10. If you are unemployed, do not stay in bed half the day and sit up watching TV half the night. Make a positive effort to structure your day, getting up and having meals at definite times. Within this structure, plan activities and interests to a regular timetable. If for one reason or another it is clear that you are not likely to get work for some time, you must not build up a picture of yourself as 'Idle and Unemployed', you must see yourself as: 'Active and Occupied'. Compare yourself with the man who has been forced to retire at 65 after a long, active working life. If he is not careful, he will slowly rot away in his bedroom slippers in front of the TV and beside the fire. If he has prepared himself for retirement, he will keep himself neat and tidy and fill his day with all those alternative activities for which he did not have time when he was working. You might well find it satisfying and useful to help in a Retired People's Club yourself.

Do not spend your time feeling sorry for yourself because you have epilepsy. Do not build up resentments because you cannot get work because of your fits. Build up a conceit of yourself, a pride in what you can do in spite of your fits. At the end of each day you must be able to look back and feel that you have done something useful.

Avoid extremes

Ordinary physical and mental effort and normal stresses are likely to keep your fits away. Excessive effort or severe stress including the stress of boredom may break down the control of your epilepsy and bring on more fits. If you have exams to take at school or university, you must plan carefully so that you work steadily and do not leave everything to a frenzy of last minute effort before the exam. You should ask for and listen to advice about what you can reasonably expect to achieve. You should talk this over with your doctor, as well as your teacher, since he will know whether your tablets are slowing you down just a little bit and so are making your teacher's expectations for you unrealistic.

Consider your family

If you are still having fairly frequent fits, you must realise that your family cannot avoid worrying that something might happen to you during an attack. If you are going out on your own, it is only fair to them to tell them roughly where you are going and when you expect to be back. Do your best to keep to this time, but, if you are delayed, let them know if you possibly can. It is easy for a bad family row to blow up when you return very late and are criticised by parents/husband/wife who have been distracted with worry for hours and take it out on you in their relief at your safe return. If you are considerate and behave reasonably and responsibly, you are much less likely to be irritated and frustrated by their efforts to look after you and protect you. Every sensible person has the right to plan his life and decide what risks to take. However, you must remember that you are not isolated on a desert island, that you are a part of a circle of family and friends whose anxieties and interests have to be taken into account.

Practical advice

Tablets

The importance of taking tablets regularly has been emphasised in Chapter 3. Most patients are very reliable but it is easy to forget a dose or be uncertain whether or not you have taken it. It is even more difficult if you have complex partial fits when at times you are muddled. It is sensible to have a small container which holds one day's tablets. Fill this every evening ready for the next day. With most drugs it should be possible for your doctor to arrange for you to take them morning and evening. If you have to have a midday dose and need to carry the container about with you, it is important that it is not breakable. A glass bottle can cause a nasty injury, if it is broken when you fall in a fit.

Incontinence

If you are unfortunate enough to be incontinent during fits, you can help by going often to the lavatory and not drinking too much before you go to bed. You should protect your mattress with a plastic sheet. If the problem is a frequent one, you can save yourself a lot of embarrassment by wearing protective pants during the daytime, and a lot of troublesome washing by using incontinence pads at night. Very occasionally special arrangements need to be made but your GP or the hospital doctor will advise you about these.

Drink

Too much fluid in the body sometimes makes fits worse and so you should not drink an unusual amount. It is best to avoid drinking any alcohol since it has an effect on the tablets you are taking. However, an occasional drink does no harm and you need not feel the odd man out when drinking a toast at a wedding. But even moderate drinking must be avoided. You must remember that you may get quite drunk on what would have little effect on someone not taking antiepileptic drugs.

Safety precautions

Although you may decide to take some risks in order to be able to live an ordinary life, you must take great care not to get burnt in a fit. Broken bones and cut heads heal but a severe burn may leave you scarred for life and add enormously to your embarrassment with other people. Open fires and hot pipes in your house must be guarded. A woman who is having frequent fits may need to give up cooking. If her attacks are worse around period time, it is sensible to arrange to have cold meals for a few days or wait till her husband comes home before working at the stove.

Privacy

For most people it is important to be private when going to the lavatory or having a bath. You should not lock the bathroom door in case you have a fit in the bath water. It is better to hang an 'engaged' notice outside. If the lavatory is small, you should get the door arranged so that it opens outwards. Otherwise, if you have a fit inside, it may be impossible for anyone to get in to help you. It is better to have a shower than a bath, but if you prefer a bath, it should be shallow and, if you have many fits, the door should be left ajar and there should be someone within earshot when you are bathing.

BEA Card/Medicalert

You should carry with you some identification which explains that you suffer from epilepsy. The British Epilepsy Association provides cards and Medicalert bracelets (see Appendix 1). Otherwise, if you have a severe fit when you are out, you are liable to be whisked away in an ambulance to the nearest hospital. Once the ambulance has been called you may not be released until the whole system has been worked through. This will give a lot of people a lot of trouble and waste a great deal of your time.

Other people with epilepsy

Almost uniquely among medical problems patients with epilepsy have been and still are considered as a group.

Therefore, what you do is bound to some extent to affect other patients and you owe it to them to reduce and not to increase the difficulties which they may have. If you kill your wife's lover and it comes out that you have fits, there will always be those who will say: 'Oh well, are you surprised; after all he's an epileptic'. On the other hand, if, despite your fits, you become a famous politician or a top tennis player, you can do a great deal of good for a great many people by admitting from your position of security that you *have* epilepsy, although you do not *suffer* from it.

7. WHAT WILL PEOPLE THINK OF ME?

Unfortunately, it must be admitted that many patients suffer more disability because of difficulties in their relations to other people than they do because of their fits. There is a great deal that can be done to prevent these difficulties arising and to do so is just as important as controlling fits.

The attitude of the public to the patient

It must be accepted that there is still considerable fear of and prejudice against the person with epilepsy. There is no point in complaining about and condemning this. It is necessary to understand why it has happened when the ordinary man in the street is naturally kindly and particularly helpful to those who are disabled.

1. There is a historical myth about epilepsy which goes back to biblical times and associates fits with supernatural powers, whether divine or demoniacal. The patient was considered as the subject of awe which contains an element of fear but more particularly of 'differentness'.
2. There is no doubt that to watch a major convulsion for the first time may be a terrifying experience. It is the

dramatic change to which it is so difficult to adapt. You can get used to a colleague who is quite severely deformed, because he is that way all the time. But there is something very frightening when a perfectly ordinary chap with whom you work and make jokes suddenly falls to the ground in a violent convulsion. It may be equally awesome when someone having a mild complex partial fit suddenly starts to behave in a strange and wholly inappropriate way. It may take an observer some time to realise that something is happening and then he may feel that it is he and not the patient who is becoming odd.

3. A considerable number of those with severe brain damage and resulting mental deficiency have generalised convulsions. The fits are so dramatically obvious and they are just the same as those of the patient without brain damage, that it is not surprising that the public, quite wrongly, may think of epilepsy in terms of the convulsing village idiot.

Much of this prejudice is due to simple ignorance about epilepsy. This is borne out by surveys which have shown that it is much less among the intelligent and better educated. For many years, epilepsy associations have done very important work in overcoming prejudice by arranging lectures and courses for key groups like teachers, doctors and social workers and in this and other countries it has been shown that such propaganda can influence opinion. Nevertheless, it is always very difficult to overcome prejudice by preaching, and in any case those who come to listen to the sermon are usually at least partly converted. It is very important that the patient, his family and his friends should not accept prejudice passively with bitter complaint, but rather actively should do their best to remove ignorance when they meet it. It is hoped that this little book will provide them with the true facts about epilepsy which will help them to do so.

The attitude of the patient to the public

The example of the family

Acquaintances, school fellows, workmates, the general public with whom the patient comes into contact, are likely to reflect to an important degree the attitudes of family and close friends to the patient and his epilepsy. If they treat him as an ordinary person and his fits as unfortunate but not disastrous, this can do a great deal to determine the way other people treat him. Unfortunately, it is still all too common for families who complain bitterly about public prejudice to show just as much prejudice themselves; to limit the patient's activities and appearances with the family because of 'what people might think if he had a fit'.

Upbringing

It does not need stating that a happy family life and a sound upbringing are the best guarantee that an adult will have a stable and well adjusted personality and will be a happy and valuable member of the community in which he lives. Should such a person have a motor car accident, suffer brain damage and later develop epilepsy, he is likely to be able to deal with the problems of his fits from the basis of a sound personality. However, when fits start in childhood, parents may have a much more formidable task in bringing up their child, and there is a real risk that, by the time he is an adult, he may have acquired secondary personality problems which will make his relations with the general public much more difficult. How can the parents help to prevent this?

All too often it is advised, even by doctors: 'Now you must not upset Willie or he will have a fit'. Such advice is disastrous. It is most important that Willie should be subject to the same discipline and if necessary punishments as the rest of the family. If he is made a special case, he will grow up thoroughly spoilt. He will use his fits as a lever for getting his own way about everything. He will play mother off against father and manipulate the whole household. If eventually someone has had enough and decides to stand up against the young tyrant, Willie is quite likely to win yet again by having a fit; real or put on. We all have to conform to the disciplines of the society in which we live. If Willie does not learn to do so with the help and love of his family, he will grow up to be a most objectionable adult whom no one wants to know. He will be rejected, but not because of his fits.

A child must be helped to do more than accept his epilepsy. He must learn to overcome any disability which it causes, to succeed despite his fits. He must make that extra effort and not use epilepsy as an excuse for making less. Again, lessons learnt as a child will mould the patient's personality for the rest of his life. Although the public is normally sympathetic to someone in difficulties, sympathy begins to wear pretty thin and to be replaced by annoyance when confronted with a perfectly healthy young man who uses the fact that he has an occasional fit as an excuse for not pulling his weight.

Most children find the passage through adolescence rather a stormy one and the struggle to be independent of family protection causes tension and friction at home. However, seldom is any permanent harm done. The adolescent patient finds this time particularly difficult and to make matters worse there may be an increase in his fits. His parents need to show a great deal of understanding if quite severe behaviour disorders are to be avoided and a difficult behaviour pattern established which lasts into adult life and adds to the patient's problems in relating to the general public. The dangers of over-protection have been considered. Many patients become lost and isolated when they leave the security and companionship of school. Unless there has been careful advance planning, they may be out of work for some time and, losing touch with their school-mates now working, their world contracts to that of their family from whom they are trying to escape. Planning for work will be dealt with in Chapter 10.

It is equally important to plan for companionship and leisure activities by joining suitable clubs and groups a year or two before leaving school. Nowadays many young people have their own motor bikes or old motor cars. A patient who is forbidden by law to drive may feel particularly isolated, frustrated and resentful. Do your best to sympathise with these feelings and tactfully help him as much as you can with his transport. You must be particularly helpful in encouraging the patient to form friendships with the opposite sex (see Ch. 9). Adolescence is the time to experiment and particularly to learn how to get on with girl friends and boy friends. You must allow him to do so, to make his own mistakes and learn from them. Otherwise he may make much more serious ones when he is older and failures will have a more lasting effect on him when you are not in the background to help. Most young people find it awkward to chew over their problems with their parents and frequently find a confidant in some family friend, teacher or clergyman. This is most useful and you must not allow yourselves to feel jealous.

Whom and how to tell

An unexpected fit invites fear and even horror in those watching, and occasionally causes chaos. It is obviously sensible that the patient or his family should explain to certain people about the possibility of attacks. Whom to tell will depend on the frequency of the fits and to some extent on the type of attack and when they happen. Some patients have fits only at night and so few people need know about them. Other attacks which often cause little disturbance include simple partial attacks which seldom go on to major convulsions, very mild complex partial fits which are unlikely to cause confusion, and infrequent petit mal absences. The more the patient can get over any unnecessary feeling of shame because he has fits, the more secure and adjusted he feels, the better will he be able to decide on the basis of ordinary common sense whom he needs to tell. Someone who is hard of hearing would not think of carrying a placard round his neck saying 'I am deaf', but if he went to a small party with people whom he did not know, he would not mind saying 'I'm afraid I'm a bit deaf, would you mind speaking up'. It is not possible to turn common sense into rules, but here are some suggestions which will be helpful.

1. School teachers should always be told that a child in their class has fits. Unless they are very rare, it avoids a lot of trouble if she explains about epilepsy to the other children. They are then likely to accept any attacks as a matter of course. If the teacher puts it over properly, they are most unlikely to tease the patient about something which he cannot help.
2. If you decide to tell someone about your fits, do not beat about the bush, and talk about 'funny turns'. Give a short simple account of what may happen, so that he is warned. Do not go to the opposite extreme and bore him with a wealth of detail. Your fits are much less important to someone else than they are to you.
3. If you have frequent fits, and unless you live somewhere where everyone knows about them, it is sensible to go about in public places with someone who can help you if you do have an attack. Such a person can cover up for you if you have a slight attack and cope if you have a big one. This may avoid you being rushed off to the nearest hospital.
4. However rare your fits *you must* warn your partner before you get married. This may sound too obvious to

mention but there are cases of wives woken up and distracted by a convulsing just-married husband.
5. If your fits are at all frequent, be thoughtful of the consequences to other people if you have one. Do not go alone into a china shop. If you hitch a lift, tell the driver what just might happen. If you do not there could be a very nasty accident if you have a fit.
6. As with so much in your relations with other people, the public will reflect your own attitude. If you are a well adjusted person and accept your epilepsy, you will find that someone you are telling about it will accept it too, as a nuisance with which you have to cope and not as a shameful disaster.

Irritability

Quite often patients, particularly those with complex partial fits, feel tensed up and irritable for from a few hours to a day or so before they have a fit. Also, it seems that some antiepileptic drugs, even in proper doses, may make patients a bit on edge. It is important that they should understand that these feelings are a part either of the build up to a fit or of their treatment. It is even more important that they should make effective efforts to control this irritability; and it can be controlled. Unlike the fit itself, it is not something outwith the patient's control. The general public will be happy to show sympathy and understanding to someone who is blind or has a broken leg, but they just will not put up with a bad tempered person who flies off the handle for no reason at all. Your fits or your treatment may be the reason for your irritability, but, if you want to get along with other people, you must make that bit of extra effort to control yourself and not try to use your epilepsy as an excuse.

Do not expect trouble

It would be foolish to suggest that there is never any prejudice against minority groups of people; whether they be Jews, or coloured, or those with epilepsy. However, the prejudice is nothing like as widespread or universal as many people in these groups imagine. Whether any individual runs up against social difficulties depends a great deal more on what sort of a person he himself is than on the group to which he belongs. It is when he is awkward, scratchy, ill-tempered and ready to take offence at next-to-nothing, that he is put in a box and called: 'A nasty Jew', or 'Another of those Epileptics'. So here is some advice to the patient:

1. Accept your epilepsy; and expect others to accept it and they nearly always will.
2. Do not moan about your epilepsy and do not talk about it all the time. It is of much less interest to other people than it is to you.
3. If you run into difficulties with other people, do not blame it on your epilepsy and public prejudice, think first whether it may not be your own fault.
4. If you behave like a thoroughly nice, ordinary sort of person, everyone else will treat you as such and will not think twice about your fits.

8. IS EPILEPSY ASSOCIATED WITH CRIME?

The myth that 'epileptics are criminals' still lingers on sufficiently to make it worth while considering the relationship between epilepsy and crime. There are still newspaper reports of crimes, often of violence, in which it is stated unnecessarily and irrelevantly that the culprit was 'epileptic'. It would be no more silly to point out that he was constipated or suffered from a peptic ulcer, but such afflictions are never mentioned.

Violence

There is no evidence that people with epilepsy are any more or less likely to commit crimes of violence than anyone else. Patients may behave violently during the confused period after a fit, particularly if people try to restrain or interfere with them. However, their violence is ill-directed and without purpose. There is no intent to harm since the patient's state of consciousness does not allow him to form the intent to do anything. On the other hand it would be stupid to pretend that patients are never violent. Like other people, they may be provoked and sometimes a thoughtless or even a malicious approach to their epilepsy may provide considerable provocation. It was explained in Chapter 7 that children who have not been brought up with proper understanding of the problems of their epilepsy may have quite serious behaviour problems when they are older. They may of course be violent. However, in both these cases violence is related quite indirectly to epilepsy.

Petty crime

A rather higher than expected number of those in prison for minor crimes have epilepsy. This is almost certainly because patients with social or employment problems become disadvantaged and slip into crime. Also this prison population will include patients with mental handicap due to brain damage (see Ch. 5) who will be more likely to get caught.

Epilepsy and apparent crime

It is not uncommon for behaviour, which is misinterpreted as criminal, to occur in relation to a fit. A patient during the period of confusion following an attack may be found wandering in someone's garden or even go into his house. Some patients who have very mild complex partial seizures, which themselves pass unnoticed, may, during the subsequent period of altered consciousness, do things which seem more or less sensible and which would be criminal if they were responsible for what they were doing. For example: a patient who went into a shop intending to buy some soap and had a mild attack might, when he was confused, pick up several packets of soap and walk out without paying for them. Occasionally during the same sort of fit a patient may start to undress or even masturbate. If this happens in public he may run into trouble with the police. Such cases should seldom cause any problem if the police use common sense and are properly advised. However, if a patient knows that he is likely to have such a type of attack fairly often, it is a sensible precaution to have someone with him when he goes about.

Epilepsy as a defence

Apart from the minor misdemeanours already mentioned, it is very rare for epilepsy to be used successfully as a defence in the case of a crime which has been investigated properly. Rather obviously from what has been written earlier in this book, to have epilepsy is no excuse, since people with epilepsy are perfectly ordinary sane people. A brain-damaged person who happens to have fits might use his mental handicap although not his epilepsy as a defence. To commit a crime it is necessary not only to do something wrong but also that it should be shown that there was the intent to do something wrong. If antisocial behaviour occurs in relation to a fit, it will be purposeless and it should be apparent that there was no intent. If, on the other hand, it can be shown that what was done was purposeful and done with intent, it could not have been done during the state of altered consciousness and irresponsibility related to a fit, and so it would be useless for the accused to claim epilepsy as his defence.

9. CAN I MARRY AND HAVE CHILDREN?

Marriage

Unless a patient is prevented by his religion from using some method of contraception, there is no medical reason why he or she should not get married. Sexual intercourse will not make fits worse. In fact lack of the opportunity for normal marital sex may build up frustrations and tensions which make fits worse. There are several points which need to be considered.

1. Each partner to a marriage has a contribution to make. Usually the husband needs to earn a wage sufficient to maintain the household, perhaps helped out by his wife's earnings. The wife, apart from paid work which she may do, has the very demanding job of looking after and organising the home. If the husband has epilepsy, he may have great difficulty in holding down a job (see Ch. 10) and this may cause financial problems or place an undue strain on the wife who needs to go out to work as well as running her home. If she does, her husband left at home, unemployed, may become resentful and lose his self-respect. If the wife has epilepsy, there are fewer problems if her fits are well controlled. However, if she is still having frequent attacks, she may be at risk if

she is doing the cooking or is left alone at home while her husband is at work.

2. As has been explained in Chapter 5, patients with brain damage may have an important degree of mental handicap as well as their fits. If there is mental subnormality in either partner, a successful marriage would be difficult from the practical point of view and patients should be persuaded not to undertake responsibilities with which they could not cope. However, it must be emphasised that their fits are not the reason for it being unwise for them to get married.

3. Some patients find it difficult to get along with others in their age group and become isolated within themselves (Ch. 5). They may run into further difficulties during the courting years of late adolescence. It is unfortunately quite common for parents to discourage boy friends or girl friends, either because they believe that the patient should not get married, or that no one should marry an 'epileptic'. The patient feels even more alone and, with a natural desire for companionship and sexual relations, may choose *any* partner; often a highly unsuitable and inadequate person who has been left on the shelf by everyone else. All too often such marriages prove to be disastrous failures and the patient's problems are increased enormously. It is helpful for family and friends to do their best to explain the possible difficulties to the patient without being too interfering.

4. It goes without saying that the idea of the big strong man who goes out hunting, or at least earns the daily bread, and the meek little woman who stays at home spinning, or at least looks after the house, is quite out of date. Nevertheless, in most happy and successful marriages there is a good deal of the 'Mr and Mrs Jack Spratt'. Each partner helps the other by doing the things which she or he does less well. The great majority of patients will gain a great deal of help and support

from a sound marriage. However, if the marriage is to be sound each must feel that he is making his contribution, as it were, in return for the support he is getting. If not, he will lose his self-respect and the marriage will fail, however well-meaning his partner may be.

5. Marriages between two people with epilepsy are perfectly sensible and may be most successful. Two patients, if they are well adjusted to their epilepsy, will have great understanding of and sympathy with each other's problems. However, the points already mentioned apply with particular force when two patients marry. The advice which they will need about having children will be considered later.

6. There is a special case which should be mentioned. A small number of patients who are unable to manage to live independently are looked after at considerable cost to the local authorities in centres for epilepsy. Many of these are sensible well adjusted people whose fits do not cause much trouble. Some form stable 'engagements' with other patients which have no chance of getting any further. It should not be beyond the imagination of Welfare Departments to help these couples with suitable housing giving some supervision and social worker support. The patients would be able to live normal lives and the authorities would be saved a great deal of money.

Having children

Before a person with epilepsy decides to have children both partners should have a talk with their GP. Before considering the question of their child having fits, they will have to decide whether it would be practical to accept the additional responsibility of looking after a family. Is the husband in work? Is his job secure? Is he earning enough money? If the husband is the patient and is unable to work and the wife is earning, how will they manage when she has to stop work to have and look after the baby? If the wife is the patient, is her epilepsy controlled well enough for her to be able to look after the baby? Unless a mother is having very frequent and uncontrolled fits, there is no reason why she should not look after her baby and it is important that she should do as much as possible for him. If he is taken away from her or she is only allowed a secondary role in his care, it is not good for the baby and the mother may develop serious emotional upsets. If the mother's fits put the baby at some risk of injury, the district nurse or health visitor will be able to advise how this can be avoided and how Granny or a good neighbour can help. However, unless a woman is going to be able to take the greater part of the responsibility for looking after *her* baby, it is better for her to postpone having one until she can.

Points to consider

A married couple will want to discuss with their GP the possibility of their child having epilepsy. There is no simple answer and each couple has to be considered as a particular case. Whether the GP advises himself or refers the couple to a specialist in these matters, the doctor will be considering these points.

1. There are some uncommon diseases which affect the brain, which cause fits, and which are hereditary. If a number of people in the patient's family has had fits and if there is any evidence of one of these diseases, the doctor will advise the couple not to have children. It is worth emphasising that there are very few cases in which the doctor will say 'No', in this way.
2. It was explained in Chapter 1 that fits occurred from a combination of a certain inherited tendency to have fits and some brain damage which might trigger them off. It follows that, if a patient has no apparent evidence of

brain damage, it is likely that he inherited a greater than average brain sensitivity or tendency to have fits. He may pass on this inherited sensitivity. On the other hand, if it is known that he has suffered important brain damage, it is likely that his fits are due to this and that there may be very little inherited tendency. For example, if a healthy married man had a serious head injury as a result of a motor car accident and later developed fits there would be only a slightly increased risk of his children having epilepsy. On the other hand, a patient had severe febrile convulsions (Ch. 1) as a small child, as did others in his family who did not develop epilepsy. When he was about 15 he started to have complex partial fits. There is some appreciable inherited brain sensitivity and this may be passed on to his children. The doctor will explain this increased risk, which the parents-to-be should understand before they decide to have a family.

3. If only one partner has epilepsy, in most cases the risk of a child having fits is not great and few doctors would advise against having a family for this reason. If, however, both partners have epilepsy, there is a much greater chance of their children being affected and most doctors would advise contraception. There is the added reason that in these cases the practical difficulties referred to are likely to be greater. Your GP will often want to refer you for expert Family Planning guidance because the commonly used 'pill' is affected by anti-epileptic drugs and slightly larger doses will be required. It is sometimes better to advise some other method.

4. If you have decided to have a family, there is an added risk that your child will have febrile convulsions. This does not mean that he will have epilepsy later on. It does mean that you should discuss the matter with your doctor and arrange to get immediate treatment for these convulsions. Most cases of epilepsy which follow febrile

convulsions are due to brain damage which has been caused by the convulsions not being treated quickly enough.
5. The decision to have children should be made by both partners after having had medical advice. If you are unlucky and do have a child with epilepsy, it is most important that you should not feel guilty or, worse still, make the partner who is a patient feel guilty. If complicated feelings of guilt develop, the family is sure to become a most unhappy one. An unhappy family with husband and wife at each other's throats will be a much greater disability to the child than his epilepsy. If you love your partner and accept his epilepsy, there is no reason at all why you should not accept it in your child.

There are three popular misunderstandings which sometimes confuse people:

1. If two people who do not have epilepsy have a child who does, in no way is this the fault of either partner, unless there is a strong family history of fits which has not been discussed with the doctor. There is no substance whatsoever in such old wives' tales as: 'His fits are due to that dreadful quarrel which I had with Aunt Agatha', or 'Of course I did tell my husband that we should not have intercourse when I was pregnant'.
2. Having a baby is NOT a cure for epilepsy. During pregnancy important changes occur in the body and sometimes these make fits better for the time being, and sometimes they make them worse. Equally often they make no difference at all. If there is a change in fit pattern during pregnancy, you should report this to your doctor since he may want to adjust the dose of your tablets.
3. If you have fits during pregnancy you are not going to produce a deformed baby. Very occasionally babies are born with some deformity. In mothers taking anti-epileptic tablets there is a very slightly increased risk of

some deformity which is usually a minor one. Doctors have been into this very carefully and are agreed that this is no reason why a patient should not have a child and certainly no reason for stopping her tablets during pregnancy. However, during pregnancy it is particularly important that the mother should not become intoxicated with her tablets (see p. 22), and that she should have the smallest dose necessary to control her attacks. Fearing that fits may harm the baby, she *must not* increase her tablets unless her doctor tells her to do so.

A happy marriage and family life is the surest foundation upon which to build a happy and worthwhile life. No person with epilepsy should be denied these opportunities without very good reason. Family and friends may help with their advice to avoid some of the possible pitfalls which might lead to a marriage disaster. However, they must remember that in our society marriage is an entirely personal matter and that, at the end of the day, it is for the patient to decide.

10. HOW CAN I GET AND KEEP A JOB?

The amount of difficulty which patients have to face in getting and keeping a job varies a great deal and depends on a number of things of which the most important are discussed below.

The frequency of attacks

Very occasionally fits prove difficult to control and patients have a great many attacks even with the best possible treatment. It is unlikely that these unfortunate people will be able to hold down an ordinary job and they should seek the help of the Social Services so that they can be placed in a suitable sheltered workshop. On the other hand there are a great many perfectly ordinary people whose epilepsy is well controlled and who have very few fits. Their problem maybe that there is some prejudice against employing someone with epilepsy. However, more often they fear or imagine that there will be such prejudice and approach employers in a resentful and suspicious way, expecting trouble.

The type of attack

Simple partial fits with little alteration of consciousness which rarely develop into generalised convulsions should cause little disability if suitable work is chosen. A person liable to have attacks of jerking of his right hand should not work with valuable or delicate objects. Petit mal absences are uncommon in those of working age, but, if they do persist into adult life, they will cause little difficulty unless they are frequent and cause periods of confusion. In this case the patient could work quite well for example as a gardener or a bookbinder but would be an unreliable accountant. Complex partial fits vary a great deal in type and severity. They may cause difficulties unless the strange behaviour which may occur during or after an attack is understood by fellow workers. Such patients should not look for work which involves contact with strangers, say as a shop assistant or a bank clerk.

When the attacks usually occur

Many patients have their attacks almost always at night, first thing in the morning, or in the evening when they are relaxed. Their fits should not interfere with their work although they should explain to their employers that after a night or early morning fit they may be late for work. Some firms who work staggered hours may be able to arrange for a patient to start and finish late.

Brain damage

It has been explained (p. 8) that brain damage often causes fits as well as a degree of mental handicap. If the brain damage is severe, attacks may be difficult to control, but the greater employment problem is likely to be the mental handicap. Many patients in this group will need to work in work centres or some sort of sheltered employment.

The effect of treatment

The patient should discuss with his GP or the hospital doctor whether the drugs which he needs to take to control his fits are likely to affect the work which he can do. They may slow him down just a little which would make it difficult for him to keep up with work in an assembly line, or cause a slight difficulty in carrying out fine finger movements which would prevent him from becoming a watch repairer. If a patient can discuss his fits with his employer and get him to accept the possibility of the occasional attack, there will be less need for him to press his doctor to increase his tablets to prevent fits at all costs. The cost of taking large doses of antiepileptic drugs is usually too high, since they interfere with a patient's ability to work properly much more than does the occasional fit.

Local conditions

Although epilepsy takes so many different forms, people tend to consider all patients together as a single group. Therefore, how each patient works and behaves tends to affect the local attitude to everyone else with epilepsy in the neighbourhood. A large firm, who employs three or four people with epilepsy, realises that they are perfectly ordinary, pleasant people and that they work as well or better than the rest, is likely to be most sympathetic when a patient applies for a job. On the other hand, one patient who has developed an awkward personality and has caused frequent disruptions at work will close the door of his factory quite effectively against any other future employee with epilepsy. It is worth remembering that epilepsy is very common and that many managers, personnel officers and foremen may have epilepsy themselves or may have close friends or family who do. If you can hear about this, you may find it much easier to get a sympathetic interview for a job.

General hints

Because of all these different factors which affect a patient's efforts to get a job, it is not possible to give individual advice but some general advice may be helpful.

Organisations which can help

The Department of Employment has trained and experienced disablement resettlement officers (DROs) whose job it is to help disabled people to find work. A patient can become registered as disabled and is given a special green card. Large firms are required by law to employ a certain percentage of such disabled people. However, patients should be a little careful about agreeing to become registered as disabled. Firstly, many people with epilepsy are not really disabled by their fits at all and, if they are registered as disabled, it implies a disability which is not there. Secondly, those patients who experience the greatest difficulty in getting work are those who in addition to their epilepsy have a degree of mental handicap or who have developed personality or behaviour problems. It is often this group who become registered and so some employers are wary of green cards saying 'epilepsy' because they have come to connect them with these associated problems rather than with epilepsy itself. The epilepsy associations (see Appendix 4) provide a very valuable advisory service and publish booklets which help to explain such matters as workmen's compensation and superannuation. These do not present the problems which some employers think.

Personal approach

It is often possible to get a lead into a job through a member of the family or a family friend. This can be particularly helpful if the lead is to an employer who is known to be sympathetic to people with epilepsy because of some personal connection. The patient, who may well have enough opposition to overcome, must not be too proud to accept such influence. Once he has got his job he can justify himself by showing that he is more than entitled to keep it. If you do get a chance of work, it is a great help to go to your doctor and ask him to write a careful letter explaining just what your fits are like and that there is no reason why they should prevent your doing the job for which you are applying. Often this approach may be much more useful than his filling in the official form necessary to get the DROs green card.

Secrecy

If you feel that you have been turned down for jobs because you have epilepsy, there is a great temptation to hide the fact when you apply for the next one. However difficult this advice may seem, if there is any reasonable chance of your having a fit during working hours, you *must* tell your employer about it. If you do not, and if you do have a fit, you may lose your job, however unfair this may seem to you. Further, if you start work terrified that you may have a fit and lose everything, you will be in such a state of tension that you will be much more likely to have an attack. Here again a personal letter from your doctor about your fits may be a great help. It is also important for either you or your doctor to explain to your employer that it is possible that, although your fits are well controlled, you may have a few attacks just at first because of the strain of starting a new job.

Training

The better a person is trained or the better his qualifications, the more likely he is to get and keep a job particularly in times of high unemployment. This applies even more so to someone who may be slightly disadvantaged by having fits. When you leave school do not be tempted into an easy-money job which does not lead anywhere. If it is at all possible, take up an apprenticeship which will give you special skills, or go to classes which will give you a qualification. Older patients may be able to take advantage of Government training schemes for the disabled which the Department of Employment will tell you about. It is important that the patient gets the best possible advice, including that of his doctor, before starting on any training. He must be assured that what he is planning to do is well within his ability and that he can complete the course with ordinary steady effort. If he is too ambitious, not only will failure add to his frustrations and lack of confidence but the strain of trying to achieve the impossible may result in the breakdown of the control of his fits.

Fellow workers

It sometimes happens that a patient gets a job through a sympathetic boss and then cannot keep it because he runs into trouble on the shop floor. Here are some suggestions to avoid this.

1. Try to find a mate to whom you can explain your fits and who will be able to help you out when you have one.
2. If your fits cause you to lose work time and this involves other people having to do your work for you, make sure that you take the opportunity of doing a bit extra to help them out. Above all do not let people think that you are using your epilepsy as an excuse for slacking or avoiding nasty jobs.
3. Do not bore people with details of your fits. You would be no more interested in incessant accounts of someone else's stomach trouble.

However, when all this has been said, it comes back to what has been emphasised through this little book. If you understand your epilepsy and have learnt to accept it, if you do not allow yourself to develop hang-ups and awkwardness of personality, if you are and behave like an ordinary decent hard-working sort of chap, other people will accept your epilepsy and will accept you for what you are. By so doing you will help not only yourself, but you will go an important part of the way to helping to remove what prejudice may remain against other people with epilepsy.

APPENDIX 1
THE BRAIN

Fits are due to a temporary disturbance in the brain. Therefore, to understand them it is helpful to have some simplified idea of how the brain works.

The brain may be thought of as made up of an immensely complex miniaturised computer (computer-brain) connected through a very elaborate telephone exchange (exchange-brain) with telephone wires (nerves) which carry messages to and from all parts of the body. There are many billions of nerves. Many of them are in the telephone system and are quite long, but the great majority are very small and make up the computer-brain.

Let us say that someone wants to get up from his chair and switch on the TV. His computer-brain prepares a programme which sends a series of instructions to the exchange-brain which in turn makes the connections necessary to send messages to the many different groups of muscles which need to be activated so that the required movements are carried out in the right order.

No one can carry out sensible actions unless he knows what is going on around him. All over the body there is a huge number of sense organs which send information along the nerves to the exchange-brain: for example touch, pain and temperature from the skin, or light and sound from the eyes and ears. Out of millions of messages, the exchange-brain selects the important ones and connects them together so that they mean something. When someone picks up an orange, his fingers send messages about its smoothness and coolness, his nose about its smell and his eyes about its orange colour. Individually, each set of messages does not mean very much. It is only when they have been sorted out in the exchange-brain that he has the information that he is holding an orange.

Every day the exchange-brain feeds millions upon millions of pieces of meaningful information to the computer-brain. Most of the workings of the computer-brain remain a mystery but probably it processes information in several ways. Some information is ignored and discarded as of little importance. Some affects the way a person feels and becomes linked up with emotions, say of fear or pleasure. Some is worth storing away in the form of memory to influence future thinking and actions. Some information needs to be dealt with at once and goes to form a programme for present action.

The computer-brain is responsible for consciousness: the awareness of what is happening and the planning of purposive action. When someone is asleep, messages are being received and sorted out by his exchange-brain but he does not appreciate them. If urgent or forceful messages are received, the computer-brain is alerted and the person wakes up and becomes conscious of what is going on. Sleep is a natural form of unconsciousness—a part of the computer-brain is resting. However, the part concerned with consciousness may be put out of action: for example, by a severe head injury or by poisonous substances which accumulate during a severe illness. The patient then becomes unconscious and in a coma from which he cannot be roused even by the most urgent messages from the exchange-brain.

Not only does a person need to be aware of what is going on around him and to take suitable action to deal with his environment, but also it is essential that his own body is working properly. A huge amount of information is received by the exchange-brain from the internal organs— the heart, the lungs, the gut and so on— and is fed into the computer-brain, processed and sent back to the exchange-brain as a programme for messages to be sent back to the organs to make any necessary adjustments. All this occurs without the person being aware of what is going on. It happens subconsciously and in fact it goes on normally even when someone is unconscious, unless he is very gravely ill. The part of the brain which deals with keeping the body going is sometimes called the primitive-brain because it is very much the same as that found in animals, the rest of whose brains has not developed to the same enormous extent as man's. The primitive-brain is of particular interest because it seems to be connected closely to that part of the computer-brain concerned with memory and emotions. The primitive-brain also has connections with the temporal lobes of the exchange-brain, so called

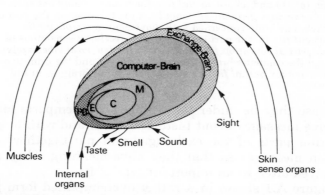

Fig. A.1 Nerves leave the exchange-brain carrying instructions to the muscles. Other nerves travel to the exchange-brain carrying information from the sense organs of the skin and from the special sense organs for taste, smell, sound and sight. The primitive-brain (P-B) receives messages from and sends instructions to the internal organs. The computer-brain includes parts for memory (M), consciousness (C), and emotions (E) whch probably lie near the primitive-brain.

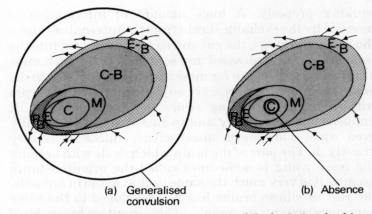

Fig. A.2 The heavy circle encloses the part of the brain involved in the fit. (a) The whole brain is involved from the beginning in a *primary generalised convulsion*. (b) Only that part of the brain concerned with consciousness is affected in a *petit mal absence*.

Fig. A.3 The heavy circle encloses the part of the brain involved in the fit. (a), (b) and (c) are *partial fits*. Each may or may not spread to give a *secondary generalised convulsion* (d). (a) arises in the part of the exchange-brain concerned with movement (a *simple partial movement fit*). (b) the part concerned with receiving sensations (a *simple partial sensation fit*). (c) arises in the *temporal lobe* involving parts of the computer-brain, the exchange-brain and the primitive-brain (*complex partial fit*). (d) involves the whole brain as in Fig. A 2(a).

because they lie under the temples. The temporal lobes receive messages about taste and smell and with neighbouring parts of the exchange-brain put together sensation messages so that they have meaning. They are involved often in an important type of fit.

Figure A.1 shows in a rather oversimplified form the connections between the sense organs, the muscles, the exchange-brain, the computer-brain and the primitive-brain.

Figures A.2 and A.3 show by a heavy circle the parts of the brain involved in the different kinds of fit described in Chapter 1. In a primary grand mal convulsion (Fig. A.2a)

72

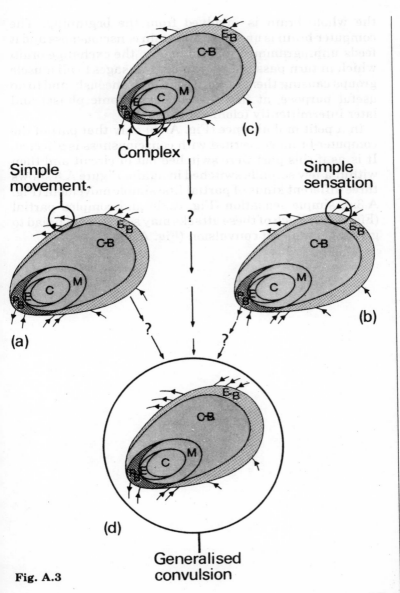

Fig. A.3

the whole brain is involved from the beginning. The computer-brain is unable to maintain consciousness and it feeds unprogrammed instructions to the exchange-brain which in turn passes disorganised messages to all muscle groups causing them to contract simultaneously and to no useful purpose, at first continuously (tonic phase) and later intermittently (clonic phase).

In a petit mal absence (Fig. A.2b) only that part of the computer-brain concerned with consciousness is affected. It is as if this part were switched out of circuit and then within a few seconds switched in again. Figure A.3 shows three different kinds of partial fits: simple movement (Fig. A.3a), simple sensation (Fig. A.3b) and complex partial (Fig. A.3c). Each of these attacks may sometimes spread to give a generalised convulsion (Fig. A.3d).

APPENDIX 2
DRIVING
LICENCES

Until 1970 no person with epilepsy could get a driving licence in Great Britain. Since 1970 new regulations have been introduced and it is now possible to get a licence provided:

1. The patient has been free of fits, while awake, for three years.
2. If he has had fits only when asleep during the past three years, he should have had fits whilst asleep but not, whilst awake, for more than three years. (This rather complicated condition is to establish that his fits occur only when he is asleep.)
3. He is not likely to be a danger to the public when driving.

If he can satisfy these conditions he can get a licence but it has to be renewed each year. However anyone who has had a fit after the age of three years is not allowed to drive a public service vehicle (bus) or a heavy goods vehicle.

Applications should be made on Form D1 which can be obtained from post offices and local taxation offices. The patient has to answer question 5e, 'Have you now, or have you ever had epilepsy?' with a 'Yes'. He will then be sent a form on which he will need to give details of his fits, the name of his doctor, and his consent to the Licensing Authority's medical adviser contacting his doctor.

If the conditions are met there should be no trouble about a licence being issued. Doubtful cases are referred to a special Medical Advisory Panel. If a patient has any difficulty in arranging for his motor insurance, he should seek help from the British Epilepsy Association (p. 78).

Having got your licence here are some practical suggestions to make sure that you keep it:

1. Do not drive when you are tired and avoid driving long distances. It is not sensible to take a job which involves a great deal of driving.
2. Be particularly careful to take your antiepileptic drugs regularly and exactly as your doctor has told you. If you miss out on your tablets and have a fit, you will have to start the three year period all over again.
3. It is particularly important that you should not take alcohol before driving.

APPENDIX 3
COMMONLY USED ANTIEPILEPTIC DRUGS

Commonly used antiepileptic drugs

Approved name	Trade name	Usual size of tablet in mg	Other sizes in mg
Acetazolamide	Diamox	250	—
Beclamide	Nydrane	500	—
Carbamazepine	Tegretol	200	100
Clonazepam	Rivotril	0.5	2
Diazepam	Valium	5	2 and 10
Ethosuximide	Zarontin / Emeside	250	—
Ethotoin	Peganone	500	—
Nitrazepam	Mogadon	5	—
Phenobarbitone	Luminal / Gardenal	60	15 and 30
Phenytoin	Epanutin / Dilantin	100	50
Primidone	Mysoline	250	—
Sodium valproate	Epilim / Depakine	200	500
Sulthiame	Ospolot	200	50
Troxidone	Tridione	300	—

APPENDIX 4
HOW TO GET HELP

Throughout the world there are a number of epilepsy associations which provide free help to patients and their families. They run information and advisory centres, organise clubs and offer holiday schemes. They arrange conferences and lectures for the general public and professional groups concerned with the problems of epilepsy such as teachers, social workers and nurses. They also raise money to sponsor research. In Great Britain you can get help from:

The British Epilepsy Association

Head Office: Crowthorne House, New Wokingham Road, Wokingham, Berkshire, RG11 3AY. Telephone Crowthorne (034 46) 3122.

Regional Offices: Third Floor, 44 Eastgate, Leeds, LS2 7JL. Telephone 0532 454416

Room 16 Claremont Street Hospital, Claremont Street, Belfast, BT9 6AQ. Telephone (0232) 40491 Ext 23

Rooms 1, 2 and 20 Guildhall Buildings, Navigation Street, Birmingham, B2 4BT. Telephone 021 643 7524

The British Epilepsy Association organises a network of voluntary groups known as Action For Epilepsy Groups which carry out their work at local level. They issue cards which patients carry, stating that they suffer from epilepsy, giving their name and address and that of their doctor and giving instructions what to do if found after a fit. They are proposing to establish at Crowthorne House a residential family advice centre and a residential career assessment unit for the school leaver.

The association also has useful literature for the patient and his family. The following are available by post from—British Epilepsy Association, New Wokingham Road, Wokingham, Berkshire, RG11 3AY.

Leaflets

For the Young Adult with Epilepsy
Epilepsy—What To Do
Facts About Fits
Driving Licences & Epilepsy
A Teacher's Guide to Epilepsy
Action for Epilepsy Campaign
Mothers with Epilepsy
Epilepsy & Alcohol
Television Epilepsy

Booklets

Schools and Centres for Epilepsy
Epilepsy and the Family
Epilepsy and Getting a Job
Employing Someone with Epilepsy

Mersey Region Epilepsy Association

Office at: 138, The Albany, Old Hall Street, Liverpool, L3 9EY. Telephone 051 236 0990

This Association runs a weekly club in Liverpool for the entertainment and companionship of the more disabled people with epilepsy, and a job creation scheme in cooperation with the Manpower Services Commission and the Social Services which employs young people with epilepsy to decorate and improve the homes of people with epilepsy, and others.

The Scottish Epilepsy Association

Office at: 48, Govan Road, Glasgow, G51 1JL. Telephone Glasgow 041 427 4911

The Scottish Association runs a work complex in Glasgow, Seaborn Industries, which comprises presently a sheltered workshop offering employment, and a work centre giving long-term industrial training. An associated occupational day centre has just been opened.

The Epilepsy Society of Edinburgh and S.E. Scotland

Office at: 13, Guthrie Street, Edinburgh, EH1 1JG. Telephone Edinburgh 031 226 5458

The Edinburgh Society runs a small hostel for women in Edinburgh which is intended as a half way stop between residential care and the community.

Medicalert bracelet or necklet

The Medicalert Foundation, 9 Hanover Street, London W1R 9HF

For a small fee this foundation provides a useful service to those liable to a medical emergency, in your case epilepsy. The patient wears either a bracelet or a necklet on which is inscribed his medical problem (epilepsy) his personal serial number and an emergency telephone number. This number may be called (the charges being reversed) by any authorised person and information provided by the patient's own doctor is given from the central file.

APPENDIX 5
GLOSSARY

The following medical terms are printed in italics at the first mention in the book. Below is a short definition of these terms. When a page reference is given it tells you where to find more detail. The words FITS, SEIZURES, and ATTACKS mean the same thing.

ABSENCE (PETIT MAL) A very mild attack with a few seconds loss of consciousness but without movements apart sometimes from a slight flickering of the eyes. Children usually have these fits. Page 2.

AURA The very early stage of *partial seizures*, before there is loss of consciousness. Therefore the patient is aware of what is happening and the aura may serve as a warning.

CLONIC PHASE That part of a *generalised convulsion (grand mal)* during which there are jerking movements of the whole body due to alternating contractions and relaxations of the muscles. Page 2.

COMA A state of deep loss of consciousness when the patient is quite unaware of what is happening and from which he cannot be roused. Page 2, 70.

COMPLEX PARTIAL FITS A very common form of seizures which usually starts in the *temporal lobe*. There is alteration of consciousness. The attacks may take many different forms. Page 4.

DÉJÀ VU A feeling that: 'what is happening has happened before'. This may be the *aura* of a *complex partial fit*.

ELECTRODES Pads applied to the head to record the *EEG*.

ELECTROENCEPHALOGRAPH (EEG) A record on paper of the electrical activity of the brain magnified many times. Page 14.

EMI SCAN A very complicated X-ray which produces a series of maps of the inside of the head to show changes in brain structure. Page 19.

ENCEPHALITIS An infection of the brain.

FEBRILE CONVULSIONS A *generalised convulsion* in an infant with a high temperature. Page 6.

GENERALISED CONVULSION A severe fit with *tonic* and *clonic phases* often followed by a period of *coma* and then sleep. Page 2.

GRAND MAL As for *generalised convulsion*.

MENINGITIS An infection of the membranes covering the brain.

PARTIAL FITS Seizures which affect only a part of the brain at first and, therefore, in which there is not complete loss of consciousness unless they spread to give a *generalised convulsion*. Page 3.

PETIT MAL See *absence*.

PRIMARY GRAND MAL A *generalised convulsion* which does not develop from a partial attack and so in which there is loss of consciousness from the beginning. Page 2.

SIMPLE PARTIAL FITS Attacks which start with a movement of or a sensation in a part of the body and in which consciousness is not lost unless the attack spreads. Page 3.

STATUS (EPILEPTICUS) A series of *generalised convulsions* without recovery of consciousness in between. A serious condition. Page 7.

TEMPORAL LOBE   That part of the brain which lies under the temple.

TEMPORAL LOBE EPILEPSY   Epilepsy starting in the *temporal lobe* usually in the form of a *complex partial fit*.

TONIC PHASE   That part of a *generalised convulsion (grand mal)* during which all the muscles contract and the body goes stiff. Page 2.

TUMOUR   A growth. Some tumours are cancers but many others are harmless.